Reviews of Environmental Contamination and Toxicology

VOLUME 199

Reviews of Environmental Contamination and Toxicology

Editor
David M. Whitacre

VOLUME 199

Springer

ISBN: 978-1-4419-3540-3 e-ISBN: 978-0-387-09808-1
DOI: 10.1007/978-0-387-09808-1

Softcover reprint of the hardcover 1st edition 2009

Printed on acid-free paper

9 8 7 6 5 4 3 2 1

springer.com

Foreword

International concern in scientific, industrial, and governmental communities over traces of xenobiotics in foods and in both abiotic and biotic environments has justified the present triumvirate of specialized publications in this field: comprehensive reviews, rapidly published research papers and progress reports, and archival documentations. These three international publications are integrated and scheduled to provide the coherency essential for nonduplicative and current progress in a field as dynamic and complex as environmental contamination and toxicology. This series is reserved exclusively for the diversified literature on "toxic" chemicals in our food, our feeds, our homes, recreational and working surroundings, our domestic animals, our wildlife and ourselves. Tremendous efforts worldwide have been mobilized to evaluate the nature, presence, magnitude, fate, and toxicology of the chemicals loosed upon the earth. Among the sequelae of this broad new emphasis is an undeniable need for an articulated set of authoritative publications, where one can find the latest important world literature produced by these emerging areas of science together with documentation of pertinent ancillary legislation.

Research directors and legislative or administrative advisers do not have the time to scan the escalating number of technical publications that may contain articles important to current responsibility. Rather, these individuals need the background provided by detailed reviews and the assurance that the latest information is made available to them, all with minimal literature searching. Similarly, the scientist assigned or attracted to a new problem is required to glean all literature pertinent to the task, to publish new developments or important new experimental details quickly, to inform others of findings that might alter their own efforts, and eventually to publish all his/her supporting data and conclusions for archival purposes. In the fields of environmental contamination and toxicology, the sum of these concerns and responsibilities is decisively addressed by the uniform, encompassing, and timely publication format of the Springer triumvirate:

Reviews of Environmental Contamination and Toxicology [Vol. 1 through 97 (1962–1986) as Residue Reviews] for detailed review articles concerned with any aspects of chemical contaminants, including pesticides, in the total environment with toxicological considerations and consequences.

Bulletin of Environmental Contamination and Toxicology (Vol. 1 in 1966) for rapid publication of short reports of significant advances and discoveries in the fields of air, soil, water, and food contamination and pollution as well as methodology and other disciplines concerned with the introduction, presence, and effects of toxicants in the total environment.

Archives of Environmental Contamination and Toxicology (Vol. 1 in 1973) for important complete articles emphasizing and describing original experimental or theoretical research work pertaining to the scientific aspects of chemical contaminants in the environment.

Manuscripts for *Reviews* and the *Archives* are in identical formats and are peer reviewed by scientists in the field for adequacy and value; manuscripts for the *Bulletin* are also reviewed, but are published by photo-offset from camera-ready copy to provide the latest results with minimum delay. The individual editors of these three publications comprise the joint Coordinating Board of Editors with referral within the Board of manuscripts submitted to one publication but deemed by major emphasis or length more suitable for one of the others.

Coordinating Board of Editors

Preface

The role of *Reviews* is to publish detailed scientific review articles on all aspects of environmental contamination and associated toxicological consequences. Such articles facilitate the often-complex task of accessing and interpreting cogent scientific data within the confines of one or more closely related research fields.

In the nearly 50 yr since *Reviews of Environmental Contamination and Toxicology* (formerly *Residue Reviews*) was first published, the number, scope and complexity of environmental pollution incidents have grown unabated. During this entire period, the emphasis has been on publishing articles that address the presence and toxicity of environmental contaminants. New research is published each yr on a myriad of environmental pollution issues facing peoples worldwide. This fact, and the routine discovery and reporting of new environmental contamination cases, creates an increasingly important function for *Reviews*.

The staggering volume of scientific literature demands remedy by which data can be synthesized and made available to readers in an abridged form. *Reviews* addresses this need and provides detailed reviews worldwide to key scientists and science or policy administrators, whether employed by government, universities or the private sector.

There is a panoply of environmental issues and concerns on which many scientists have focused their research in past yr. The scope of this list is quite broad, encompassing environmental events globally that affect marine and terrestrial ecosystems; biotic and abiotic environments; impacts on plants, humans and wildlife; and pollutants, both chemical and radioactive; as well as the ravages of environmental disease in virtually all environmental media (soil, water, air). New or enhanced safety and environmental concerns have emerged in the last decade to be added to incidents covered by the media, studied by scientists, and addressed by governmental and private institutions. Among these are events so striking that they are creating a paradigm shift. Two in particular are at the center of ever-increasing media as well as scientific attention: bioterrorism and global warming. Unfortunately, these very worrisome issues are now super-imposed on the already extensive list of ongoing environmental challenges.

The ultimate role of publishing scientific research is to enhance understanding of the environment in ways that allow the public to be better informed. The term "informed public" as used by Thomas Jefferson in the age of enlightenment conveyed the thought of soundness and good judgment. In the modern sense, being "well informed" has the narrower meaning of having access to sufficient information. Because the public still gets most of its information on science and technology from TV news and reports, the role for scientists as interpreters and brokers of scientific information to the public will grow rather than diminish.

Environmentalism is the newest global political force, resulting in the emergence of multi-national consortia to control pollution and the evolution of the environmental ethic. Will the new politics of the 21st century involve a consortium of technologists and environmentalists, or a progressive confrontation? These matters are of genuine concern to governmental agencies and legislative bodies around the world.

For those who make the decisions about how our planet is managed, there is an ongoing need for continual surveillance and intelligent controls, to avoid endangering the environment, public health, and wildlife. Ensuring safety-in-use of the many chemicals involved in our highly industrialized culture is a dynamic challenge, for the old, established materials are continually being displaced by newly developed molecules more acceptable to federal and state regulatory agencies, public health officials, and environmentalists.

Reviews publishes synoptic articles designed to treat the presence, fate, and, if possible, the safety of xenobiotics in any segment of the environment. These reviews can either be general or specific, but properly lie in the domains of analytical chemistry and its methodology, biochemistry, human and animal medicine, legislation, pharmacology, physiology, toxicology and regulation. Certain affairs in food technology concerned specifically with pesticide and other food-additive problems may also be appropriate.

Because manuscripts are published in the order in which they are received in final form, it may seem that some important aspects have been neglected at times. However, these apparent omissions are recognized, and pertinent manuscripts are likely in preparation or planned. The field is so very large and the interests in it are so varied that the Editor and the Editorial Board earnestly solicit authors and suggestions of underrepresented topics to make this international book series yet more useful and worthwhile.

Justification for the preparation of any review for this book series is that it deals with some aspect of the many real problems arising from the presence of foreign chemicals in our surroundings. Thus, manuscripts may encompass case studies from any country. Food additives, including pesticides, or their metabolites that may persist into human food and animal feeds are within this scope. Additionally, chemical contamination in any manner of air, water, soil, or plant or animal life is within these objectives and their purview.

Manuscripts are often contributed by invitation. However, nominations for new topics or topics in areas that are rapidly advancing are welcome. Preliminary communication with the Editor is recommended before volunteered review manuscripts are submitted.

Summerfield, North Carolina D.M.W.

Contents

Chapter 1
Assessing the Discharge of Pharmaceuticals Along the Dutch Coast of the North Sea

N. Walraven and R.W.P.M. Laane

Contents

1 Introduction

The presence of pharmaceutical chemicals in the environment was mentioned in the late 1970s by Highnite and Azarnoff (1977), and in the mid-1980s by Richardson and Bowron (1985) and Rogers et al. (1986). However, little attention was paid to these substances as potential environmental pollutants until the early 1990s (Weigel 2003). Stan and Linkerhägner (1992) were among the first to identify high concentrations of clofibric acid, a metabolite of the lipid-regulating agents of the pharmaceuticals clofibrate and etofibrate in groundwater. Several research groups studied and reported the presence of a vast array of pharmaceutical residues, often called PPCPs (pharmaceutical and personal care products) (Carbella et al. 2007), in municipal wastewater, surface water, groundwater, and drinking water at nanogram per liter (ng L^{-1}) to microgram per liter (μg L^{-1}) concentration levels (Stan et al. 1994; Hirsch et al. 1996, 1999; Stan and Heberer 1996; Stumpf et al. 1996; Buser

N. Walraven and R.W.P.M. Laane (✉)
DELTARES/UvA,
2600 MH Delft, The Netherlands
E-mail: remi.laane@deltares

D.M. Whitacre (ed.) *Reviews of Environmental Contamination and Toxicology* Volume 199, doi: 10.1007/978-0-387-09808-1_1, © Springer Science+Business Media, LLC 2008 et al. 1998a,b, 1999; Hartmann et al. 1998; Ternes 1998; Hartig et al. 1999; Ternes and Hirsch 2000; Heberer 2002; Tixier et al. 2003; Weigel et al. 2004; Comoretto and Chiron 2005).

To identify the presence of human and veterinary pharmaceuticals in Dutch surface-, drinking-, sewage-, and industrial water, extensive monitoring campaigns were conducted between 2000 and 2003 (Schrap et al. 2003; Versteegh et al. 2003; Mons et al. 2000, 2003; Sacher and Stoks (2003). During these campaigns ~100 active ingredients (AIs) were selected for monitoring from about 850 human and 200 veterinary pharmaceuticals used in the Netherlands. These AIs were selected on the basis of having (1) analytical techniques adequate to identify and quantify them, (2) concentrations previously reported in other studies, and (3) risk to the aquatic environment (Schrap et al. 2003). These monitoring studies emphasized analysis of the AIs rather than of their metabolites. The categories of pharmaceuticals selected, included antibiotics, coccidiostatics, analgesics, X-ray contrast media, drugs used to treat coronary vascular disease, antineoplastic drugs, and antiepileptics.

In general, concentrations of pharmaceuticals in Dutch surface water ranged from a few to several hundred ng L^{-1}. Sewage water from a residential area contained household analgesics, lipid-lowering agents, beta blockers, and antiepileptics, in concentrations up to tens of μg L^{-1} (Schrap et al. 2003). The authors found that concentrations of pharmaceuticals in municipal sewage water, and in influent water at sewage treatment plants (STPs), are very similar to those in sewage water effluents from residential areas. STP effluents, however, contained higher concentrations of specific antibiotics and X-ray contrast media. Concentrations of pharmaceuticals in surface waters were significantly lower than those in sewage water (Schrap et al. 2003). Substances most frequently identified in Dutch surface water (in at least 50% of sampled locations), during the monitoring campaigns described above, were X-ray contrast media, four analgesics (acetylsalicylic acid, diclofenac, ibuprofen, and naproxen), two antiepileptics (carbamazepine and primidone), two beta blockers (sotalol and atenolol), two antibiotics (azitromycin and sulfamethoxazole), and one anesthetic (lidocaine) (Schrap et al. 2003). Pharmaceuticals were detected only incidentally in Dutch ground- and drinking-water studies (Mons et al. 2003; Vethaak et al. 2002).

Some pharmaceuticals present in Dutch sewage and surface water ultimately end up in the North Sea. Unfortunately, concentrations of pharmaceuticals that reach seawater, along the Dutch coastal zone (DCZ) of the North Sea, have not been measured. However, in other studies, clofibric acid (Buser et al. 1998a; Weigel 2003) and caffeine (Weigel 2003) were detected at ng L^{-1} levels in North Sea water. Concentrations of diclofenac, ibuprofen, propyphenazone, and ketoprofen were also measured, but were not unambiguously identified in the North Sea water samples (Weigel 2003). In summary, the number of studies that reported concentrations of pharmaceuticals in the North Sea is limited ($n = 6$); hence, it is probable that pharmaceuticals (other than clofibric acid) exist in seawater of the North Sea (German Bight, English Coast, and Norwegian Coast), and will eventually be detected.

Because pharmaceuticals are biologically active, are designed to induce pharmacological effects at low concentrations, and are present in seawater, it is possible that they affect nontarget organisms in the aquatic environment (Rijs et al. 2003). Once emission sources and fluxes of pharmaceuticals to the DCZ are known and quantified, it will be possible to design contamination prevention measures for specific (groups of) pharmaceuticals.

In this chapter, we assess the input of both human and veterinary pharmaceuticals along the DCZ. Initially, we collected and assembled all public data on pharmaceuticals known to exist in the Dutch aquatic environment that may end up in the water of the DCZ. Then we performed emissions calculations to determine the riverine input and direct discharge (point source discharge at the end of a pipe into an aquatic environment) of circa 100 active pharmaceutical agents to the North Sea.

The sources, presence, and impact of pseudo hormones and hormones in the Dutch aquatic environment are published elsewhere (Belfroid et al. 1999; Vethaak et al. 2002).

2 Methods: Data Inventory and Emissions Calculations

2.1 Data Inventory

An inventory was assembled on existing pharmaceutical concentrations in surface-, sewage-, industrial-, and marine-water in the Netherlands; the inventory was sourced from scientific publications, and from other sources, mainly "grey" literature and unpublished data, from research institutes, for example, Netherlands Organization for Applied Scientific Research (TNO), Centre for Water Management (WD), National Institute for Public Health and the Environment (RIVM) and Applied Research for Water Management (STOWA), and universities (e.g., University of Utrecht). In addition to securing information through direct contacts with these institutes and universities, a literature search was performed on the internet and in the databases of university libraries.

From these inventoried studies, it appeared that there are several routes by which human and veterinary pharmaceuticals can enter the environment. The three most important routes are (1) direct disposal at manufacturing sites, (2) excretion via urine and feces (sewage water), and (3) dispersion or runoff of manure, etc. from agricultural land (e.g., Mons et al. 2000).

Human pharmaceuticals are primarily discharged in wastewater from households, hospitals, and nursing homes, from excreta (urine and feces) (Mons et al. 2000; Derksen et al. 2001). This wastewater enters the sewage system and is treated in a STP. From a quantitative point of view, excretion of consumed pharmaceuticals is much more significant than direct disposal from manufacturing activities (Derksen et al. 2001). Only a small percentage (1–5%) of pharmaceuticals reach wastewater from manufacturing processes (Mons et al. 2000). Evidence indicates that Dutch households directly dispose of only a small proportion of pharmaceuticals to wastewaters. In the Netherlands, 8.3% of all prescribed medicines are never consumed, and only 3% of these end up in the sewer (Blom et al. 1995). The majority of unused medicines are returned to pharmacies. Nevertheless, one-third (33%) of liquid medicines returned to pharmacies end up in the sewer (Blom et al. 1995).

The fate of excreted veterinary drugs differs considerably from the fate of those excreted by humans (Mons et al. 2000). In general, municipal sewage and, therefore, human pharmaceuticals will pass through a STP prior to entering rivers, streams, or the coastal zone. Veterinary drugs are more likely to contaminate soil and ground water (without previous waste water treatment), when liquid manure is applied to the soil as fertilizer. It is assumed that veterinary pharmaceuticals will reach surface waters only after runoff (Mons et al. 2000).

Other means by which pharmaceuticals reach the coastal and marine environment are atmospheric deposition and sea-shipping emissions. The emission of pharmaceuticals to the atmosphere is mentioned in one study only (Hamscher et al. 2002). The exact extent (size) of this source is unknown, but Boxall et al. (2003) estimated that this source is of minor importance compared to other environmental input sources. Therefore, atmospheric deposition, as an input source of pharmaceuticals to the DCZ, is not included in this study.

Although some pharmaceuticals (e.g., the antibiotic tetracycline) are mentioned as possible antifouling paint additives (Peterson et al. 1993), a literature search by Klijnstra (2005) found no commercially available antifouling products that contained pharmaceuticals. Hence, sea-shipping emissions, as input source of pharmaceuticals to water of the DCZ, are not considered in this study.

2.2 Load Calculations

The load of a specific pharmaceutical and/or metabolite transported by rivers, sewage effluent, or industrial effluent to the Netherlands (from neighboring countries) and to the DCZ is calculated according to the following equations:

$$L-\mathrm{sw}_{\mathrm{in}} = (Q_{\mathrm{a}_{\mathrm{EYS}}} \times C_{\mathrm{EYS}} + Q_{\mathrm{a}_{\mathrm{LOB}}} \times C_{\mathrm{LOB}} + Q_{\mathrm{a}_{\mathrm{SOD}}} \times C_{\mathrm{SOD}}) \times f_{\mathrm{in}} / 10^{12} \quad (1)$$

$$L-\mathrm{sw}_{\mathrm{out}} = (Q_{\mathrm{a}_{\mathrm{AND}}} \times C_{\mathrm{AND}} + Q_{\mathrm{a}_{\mathrm{MAA}}} \times C_{\mathrm{MAA}} + Q_{\mathrm{a}_{\mathrm{HAR}}} \times C_{\mathrm{HAR}}) \times f_{\mathrm{out}} / 10^{12} \quad (2)$$

$$L-\mathrm{sew}_{\mathrm{out}} = Q_{\mathrm{a}_{\mathrm{SEW}}} \times C_{\mathrm{SEW}} / 10^{12} \quad (3)$$

$$L-\mathrm{ind}_{\mathrm{out}} = Q_{\mathrm{a}_{\mathrm{IND}}} \times C_{\mathrm{IND}} / 10^{12} \quad (4)$$

where

L-$\mathrm{sw}_{\mathrm{in}}$ = Load of pharmaceuticals entering the Netherlands with surface water in 2002 (t yr^{-1}).

Fig. 1 Sampling locations (•) in the Netherlands. *EYS* Eijsden, *LOB* Lobith, *SOD* Schaar van Ouden Doel, *AND* Andijk-Ysselmeer, *MAA* Maassluis, *HAR* Haringvliet

$L\text{-sw}_{out}$ = Load of pharmaceuticals entering the DCZ with surface water in 2002 (t yr^{-1}).

$L\text{-sew}_{out}$ = Load of pharmaceuticals entering the DCZ with sewage water in 2002 (t yr^{-1})

$L\text{-ind}_{out}$ = Load of pharmaceuticals entering the DCZ with industrial water in 2002 (t yr^{-1}).

Q_a = Average flow rate at sample location in 2002 (10^9 m^3 yr^{-1}).

C = Average concentration of a pharmaceutical in surface water (SW), sewage water (SEW) or industrial water (IND) in 2002 (ng L^{-1}).

f = Correction factor (corrects for loads in unsampled surface waters entering the Netherlands and DCZ).

The various sampling locations for pharmaceutical analyses are pictured in Fig. 1.

In this study, the median loads of individual pharmaceuticals are all calculated for sampling yr 2002. Values below the limit of detection (LOD) are replaced with zero to get a conservative estimate of pharmaceutical loads.

The average flow rate (Q_a) of water in the rivers at the sampling locations Eysden (EYS), Lobith (LOB), and Schaar van Ouden Doel (SOD) (Fig. 1) are 10.925×10^9 m^3 yr^{-1}, 93.793×10^9 m^3 yr^{-1}, and 2.735×10^9 m^3 yr^{-1}, respectively (data extracted from Waterbase, 2006). The contribution of other inflow sources in 2002 has been estimated at 0.504×10^9 m^3 yr^{-1} (Netherlands Hydrological Society 2004). The average flow rate at the sampling locations Andijk (AND), Maassluis (MAA), and Haringvliet (HAR) (Fig. 1) are 19.856×10^9 m^3 yr^{-1}, 50.188×10^9 m^3 yr^{-1}, and 30.332×10^9 m^3 yr^{-1}, respectively. The contribution of other outflow sources has been estimated at 9.795×10^9 m^3 yr^{-1} (OSPAR 2004). The annual average flowrates of sewage water equates 0.168×10^9 m^3 yr^{-1}, and for industrial water the corresponding amount is 0.002×10^9 m^3 yr^{-1} (OSPAR 2004).

3 Results: Pharmaceutical Emissions in the Netherlands

The constructed database comprises over 100 pharmaceuticals of both human and veterinary origin; pharmaceuticals were measured in surface, sewage, and industrial water from the Netherlands in the period between 1996 and 2005. Most analyses were performed on surface water samples (n = 634); only a limited number of samples of sewage water (n = 13) and industrial water (n = 10) were analyzed. It should be noted that the measurements on industrial waters are highly biased, because primarily effluent water from industries that discharge large quantities of pharmaceuticals (e.g., pharmaceutical companies and hospitals) were sampled.

3.1 Occurrence and Loads of Pharmaceuticals in Surface-, Sewage-, and Industrial Water in the Netherlands

Concentrations greater than the LOD of pharmaceuticals in surface-, sewage-, and industrial-water from the Netherlands are visualized in box-whisker plots in Figs. 2–4, respectively. In these figures, seven pharmaceutical groups are distinguished: (1) antibiotics, (2) veterinary antibiotics, (3) analgesics/antipyretics/anti-inflammatory drugs, (4) X-ray contrast media, (5) fibrates/lipid regulators, (6) beta blockers, and (7) others (anthelmentics, bronchitis/asthma drugs, antineoplastics, antioestrogen drugs, neuroleptics, anesthetics, steroids, and antifungal veterinary drugs).

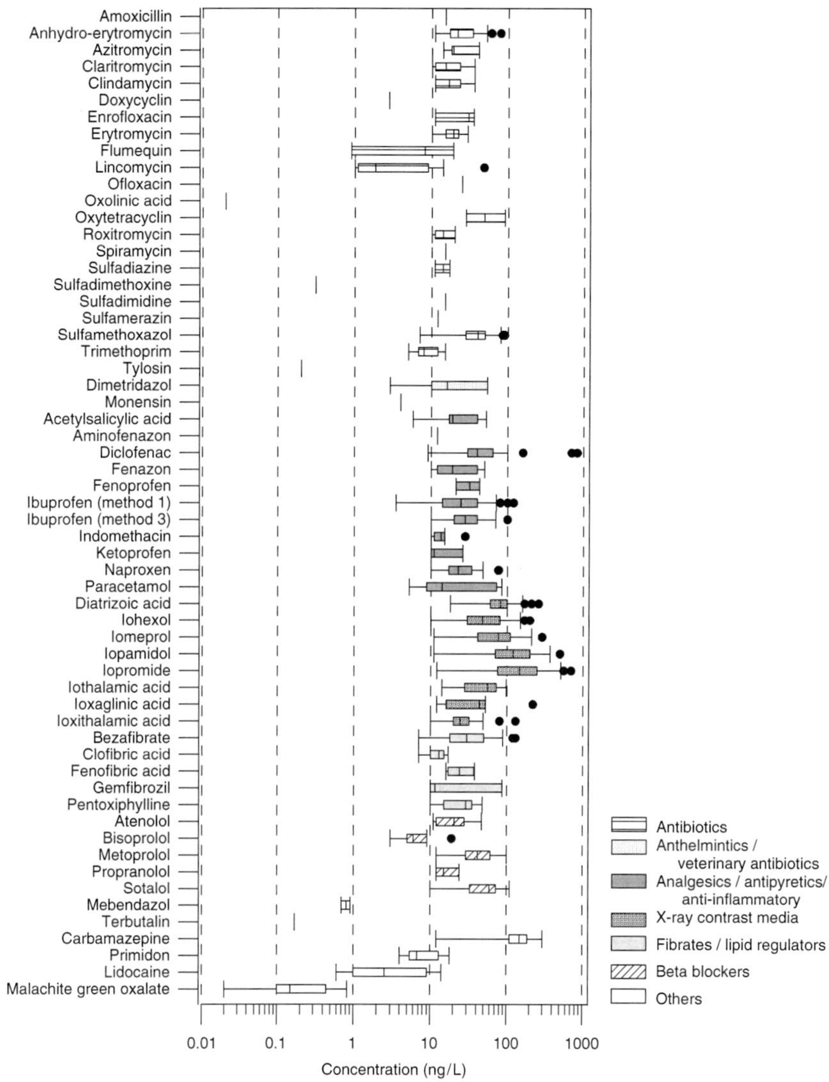

Fig. 2 Box-whisker plot of pharmaceutical concentrations in surface water of the Netherlands between 1996 and 2005 (• outlier)

Fig. 2 shows that only 58 of the 102 pharmaceuticals listed in the database have, in one or more surface-water samples, concentrations above the LOD. Concentrations of pharmaceuticals in Dutch surface water varied from 0.02 ng L^{-1} (oxolinic acid in surface water from Nieuwersluis) to 830 ng L^{-1} (diclofenac in surface water from the river Meuse at Roosteren). The highest median concentrations in surface water were observed for X-ray contrast media, with median concentrations ranging from 25 to 145 ng L^{-1}. The pharmaceuticals with the highest median concentration of each

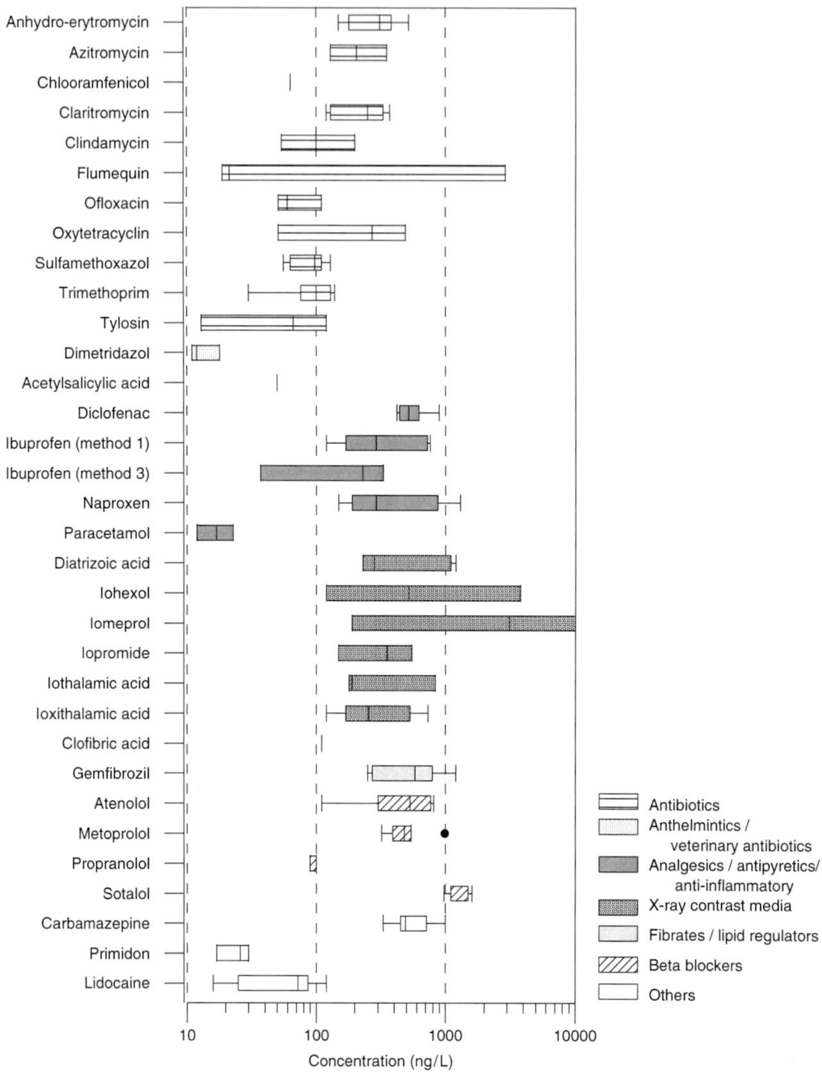

Fig. 3 Box-whisker plot of pharmaceutical concentrations in sewage water effluents from the Netherlands in 2002

of the seven above-mentioned pharmaceutical groups were oxytetracyclin (49 ng L^{-1}), dimetridazol (16 ng L^{-1}), diclofenac (40 ng L^{-1}), iopromide (145 ng L^{-1}), bezafibrate (30 ng L^{-1}), sotalol (60 ng L^{-1}), and carbamazepine (150 ng L^{-1}). The observed concentrations of pharmaceuticals in the Netherlands are in agreement with the concentrations measured in the river Rhine (Basel, Düsseldorf, and Cologne) in 2000 (Weil and Knepper 2006).

Probably, no uniform pattern exists in the geographical distribution of the mean pharmaceutical concentrations, because only a limited number of samples were

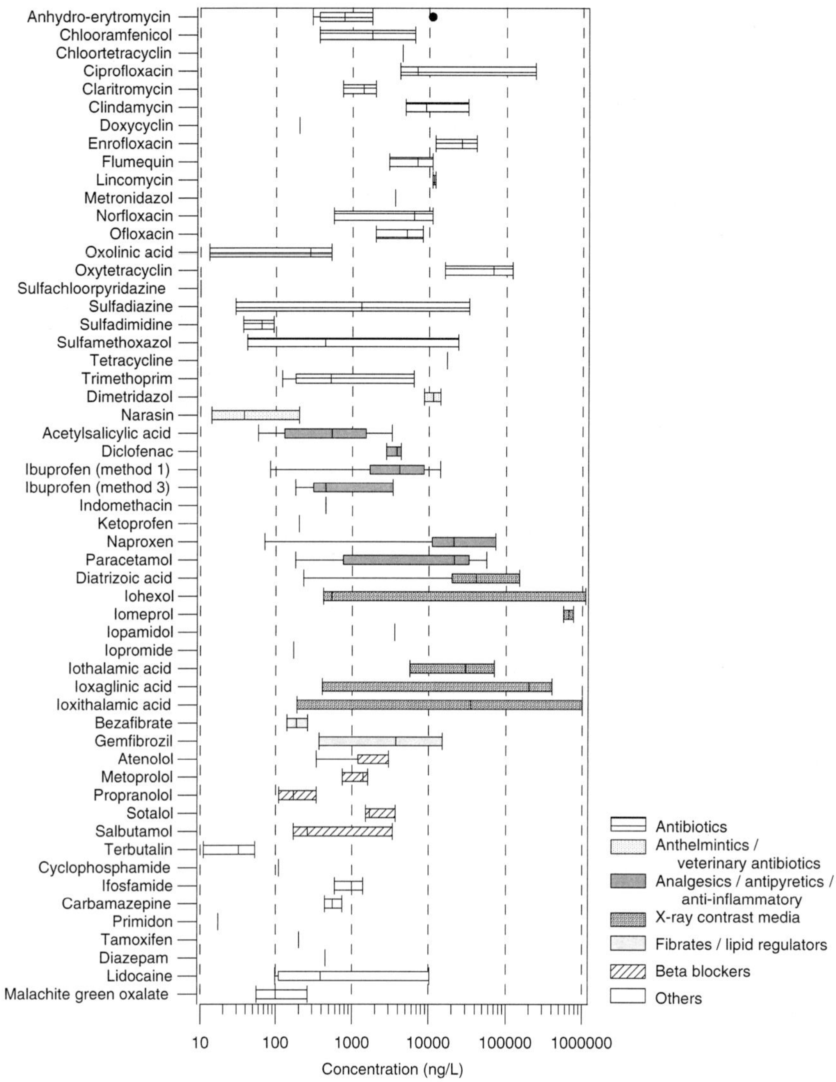

Fig. 4 Box-whisker plot of pharmaceutical concentrations in industrial waste water from the Netherlands in 2002 (• outlier)

taken and analyzed at each location. In addition, the sampling date of each location varied considerably. The only uniform patterns observed were the relatively low concentrations of pharmaceuticals in the surface waters of the northern provinces, in comparison with the more southern provinces. However, it was observed that there were fewer surface-water samples from northern provinces than from southern provinces. In general, the highest concentrations occurred in the main rivers Rhine (LOB), Meuse (EYS and MAA), Schelde (SOD), and Amsterdam–Rijnkanaal (NGN); in the Netherlands, these are situated in the southern provinces. Iopromide,

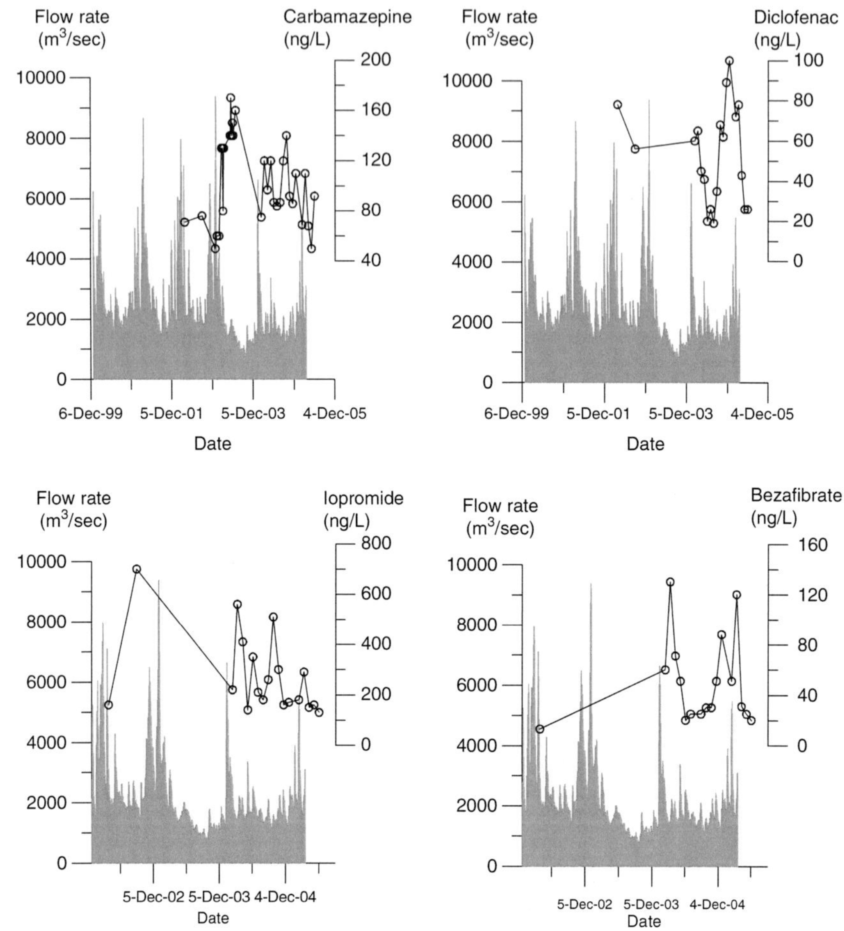

Fig. 5 Flow rate and concentrations of carbamazepine, diclofenac, iopromide, and bezafibrate versus time in the Rhine near Lobith

carbamazepine, and sotalol were present (concentrations > LOD) at almost all sampled locations.

A nonsignificant relation was found between the flow rate and the concentrations of carbamazepine, diclofenac, iopromide, and bezafibrate in the river Rhine near Lobith for the period 2001–2005 (Fig. 5). This suggests that other important variables exist that influence pharmaceutical concentrations in surface water. These variables probably constitute the application period and amounts of pharmaceuticals discharged to the environment. Pharmaceuticals are, in contrast to some other chemical substances (e.g., nitrate and phosphate), discharged throughout the yr. However, discharge of some pharmaceuticals (e.g., analgesics) is higher during winter than in summer. This may explain the lack of a relationship between flow rate and diclofenac concentrations in surface waters near Lobith.

Pharmaceutical concentrations in sewage-water effluents are generally higher than in surface water. Fig. 3 shows that only 32 of the 102 pharmaceuticals in the database have, in one or more sewage-water effluent samples, concentrations above the LOD. Concentrations of pharmaceuticals in Dutch sewage-water effluents varied from 11 ng L^{-1} (dimetridazol) to 10 μg L^{-1} (iomeprol). The highest median concentrations in sewage-water effluents were observed for X-ray contrast media, with median concentrations ranging from 190 to 3,100 ng L^{-1}. The pharmaceuticals with the highest median concentration of each of the seven above-mentioned pharmaceutical groups were anhydro-erytromycin (310 ng L^{-1}), dimetridazol (12 ng L^{-1}), diclofenac (520 ng L^{-1}), iomeprol (3.1 μg L^{-1}), gemfibrozil (580 ng L^{-1}), sotalol (1.1 μg L^{-1}), and carbamazepine (490 ng L^{-1}).

Pharmaceutical concentrations in industrial wastewater effluents are generally higher than in surface or sewage-water effluents. Most industrial wastewater samples are collected from pharmaceutical industry facilities and hospitals. Fig. 4 shows that only 54 of the 102 listed pharmaceuticals in the database have, in one or more industrial wastewater effluent samples, concentrations above the LOD. Concentrations of pharmaceuticals in Dutch industrial wastewater effluents varied from 11 ng L^{-1} (terbutalin) to 1.1 mg L^{-1} (iohexol). The highest median concentrations in industrial wastewater effluents were observed for the X-ray contrast media, with median concentrations ranging from 540 ng L^{-1} to 665 μg L^{-1}. The pharmaceuticals with the highest median concentration of each of the seven pharmaceutical groups were: oxytetracyclin (68 μg L^{-1}), dimetridazol (11.3 μg L^{-1}), paracetamol (21.4 μg L^{-1}), iomeprol (665 μg L^{-1}), gemfibrozil (3.73 μg L^{-1}), sotalol (1.7 μg L^{-1}), and ifosfamide (995 ng L^{-1}).

The pharmaceuticals discharged to the DCZ are dominated by X-ray contrast media ($n = 7$), followed by antibiotics ($n = 6$), analgesics/antipyretics/anti-inflammatory drugs ($n = 2$), beta blockers ($n = 2$), fibrates/lipid regulators ($n = 1$), veterinary antibiotics ($n = 1$), and others ($n = 1$).

Although the concentrations of many pharmaceuticals in sewage- and industrial-water effluents are higher than those in surface water, the estimated loads of pharmaceuticals directly discharged to the DCZ are low compared to the riverine input to the DCZ. This is a result of the huge difference in flow rates between surface waters and sewage and industrial water entering the DCZ. The direct discharge loads of the top 20 pharmaceuticals are shown in Table 1. The highest pharmaceutical loads discharged to the DCZ in sewage-water effluents and industrial water were calculated for sotalol to be from 0.16 to 0.27 t yr^{-1}, with a median of 0.18 t yr^{-1}, and for diatrizoic acid to range from 0 to 0.327 t yr^{-1}, with a median of 0.067 t yr^{-1}. The direct discharge of pharmaceuticals to the DCZ was calculated to be <5% of the corresponding riverine discharge of pharmaceuticals.

The median loads of the top 20 pharmaceuticals entering the Netherlands by rivers Rhine, Meuse, and Scheldt are presented in Table 1. The median loads vary from 0.05 t yr^{-1}, for lincomycin, to 42.7 t yr^{-1}, for iopromide. The highest loads result from X-ray contrast media, with values up to 42.7 t yr^{-1} (iopromide). Barreveld et al. (2001) estimated the median load of carbamazepine entering the Netherlands, near Lobith in 1997, to be 4.1 t yr^{-1}, with an average of 6.3 t yr^{-1}. These values agree well with the loads calculated in this study for carbamazepine (total of 7.5 t yr^{-1}) in 2002.

Table 1 Balance (input-total discharge) between the annual load of pharmaceuticals entering the Netherlands by rivers in 2002 (riverine input) and the annual load leaving the Netherlands in 2002 (riverine and direct discharge). The total sales of pharmaceuticals in the Netherlands in 1999 (Derksen et al. 2001) is also indicated

	Riverine discharge	Direct discharge (t/yr)		Total discharge	Riverine input	Input-total discharge	Sales (1999)
Pharmaceutical	(t/yr)	STP	Industrial	(t/yr)	(t/yr)	(t/yr)	(t)
Iopromide	22.611	0.000	0.000	22.611	42.726	20.115	
Carbamazepine	12.064	0.082	0.001	12.147	7.467	−4.680	10.8
Diatrizoic acid	8.989	0.047	0.067	9.103	15.261	6.158	
Iopamidol	8.841	0.000	0.000	8.841	15.608	6.767	
Iomeprol	6.418	0.032	0.001	6.451	7.442	0.992	
Sulfamethoxazol	5.245	0.016	0.000	5.261	2.745	−2.516	4.1
Iohexol	4.637	0.020	0.001	4.658	6.775	2.117	
Sotalol	4.631	0.185	0.003	4.819	9.050	4.231	
Diclofenac	3.984	0.087	0.003	4.075	10.545	6.470	
Metoprolol	2.808	0.081	0.001	2.890	2.530	−0.360	
Anhydro-erytromycin	2.534	0.052	0.001	2.587	2.205	−0.382	
Ibuprofen	2.341	0.049	0.009	2.399	1.344	−1.055	48.0
Azitromycin	2.093	0.034	0.000	2.128	2.200	0.073	
Dimetridazol	2.066	0.001	0.000	2.067	0.835	−1.232	
Iothalamic acid	1.845	0.030	0.006	1.882	3.312	1.430	
Lincomycin	1.322	0.000	0.000	1.322	0.053	−1.269	
Bezafibrate	1.262	0.000	0.000	1.263	0.920	−0.343	0.3
Claritromycin	1.212	0.042	0.000	1.254	1.199	−0.055	
Roxitromycin	1.102	0.000	0.000	1.102	1.276	0.174	
Ioxithalamic acid	0.895	0.032	0.013	0.940	2.785	1.845	

STP sewage treatment plant

3.2 *Mass Balance of Pharmaceutical Loads*

The difference between the load of pharmaceuticals entering the Netherlands by rivers and leaving by rivers, together with the amounts directly discharged in the Netherlands, provides a net balance of pharmaceuticals that are added or removed from Dutch surface water (Table 1).

It is remarkable that amounts of the top 20 X-ray-contrast-media agents entering the Netherlands exceed amounts leaving the Netherlands. Amounts of X-ray contrast media removed from surface water in the Netherlands vary (at least) from 1 t yr^{-1}, for iomeprol, to 20 t yr^{-1}, for iopromide. These amounts are probably not removed from surface water by STPs. Schrap et al. (2003) determined that removal rates for X-ray contrast media by STPs in the Netherlands are less than 10%. Ternes and Hirsch (2000) found that no X-ray contrast medium was removed in German STPs. How X-ray contrast media are removed from surface water is not yet known. Possibly, they adsorb to suspended matter and are deposited in the river sediment.

Schrap et al. (2003) report removal efficiencies of 8-12%, 48-67%, and 6-63% for sotalol, diclofenac, and azitromycin, respectively. No data on removal rates of roxitromycin, during sewage-water treatment, were found in literature.

All other pharmaceuticals listed in Table 1 show a net addition of pharmaceuticals in the Netherlands. The amount of bezafibrate sold in the Netherlands, in 1999 (Derksen et al. 2001), agrees with the net addition of bezafibrate in 2002 (Table 1).

Table 1 shows that analgesics are by far the most highly consumed (sold) pharmaceuticals in 1999. However, the only analgesics in the top 20 (Table 1) are diclofenac and ibuprofen. Paracetamol does not appear among the top 20, although 223 t was sold in the Netherlands in 1999. The low discharge rate of paracetamol to the DCZ (0.32 t yr^{-1} in 2002) is caused by the high removal efficiency of paracetamol in Dutch STPs. Schrap et al. (2003) reported 100% removal percentages for paracetamol. Fent et al. (2006) also reported total removal of paracetamol by STPs. Although less ibuprofen and diclofenac are sold than paracetamol (Table 1), more of these pharmaceuticals are discharged to the DCZ as a result of the lower STP removal percentage of these pharmaceuticals (52–96% and 48-67% for ibuprofen and diclofenac, respectively) (Schrap et al. 2003).

3.3 Pharmaceuticals Versus OSPAR Priority Substances

Mandatory monitoring of the following chemical substances was initiated in 2002 as a result of the framework of the OSPAR convention for the protection of the marine environment of the North Sea: total Hg (mercury), total Cd (cadmium), total Cu (copper), total Zn (zinc), total Pb (lead), lindane, ammonia, expressed as N, nitrates, expressed as N, orthophosphates, expressed as P, total N, and total P. PCBs (polychlorinated biphenyls) were measured on a voluntary basis. The discharge data of ten pharmaceuticals to the DCZ in 2002, as estimated in this study, are together with the OSPAR priority substances presented in Table 2.

Table 2 shows that the discharge of the pharmaceuticals iopromide, carbamazepine, diatrizoic acid, iopamidol, iomeprol, syulfamethoxazol, sotalol, iohexol, diclofenac, and metoprolol to the DCZ is in the same order of magnitude as the discharge of the mandatory OSPAR substances Cd and Hg. The discharge of the listed pharmaceuticals is an order of magnitude larger than that of the recommended OSPAR substance PCBs, and two orders of magnitude larger than lindane (mandatory monitoring).

3.4 Environmental Impact of Discharged Pharmaceuticals

Among pharmaceuticals discharged in significant amounts to the DCZ are those that have human and ecotoxicological risks that are largely unknown (Jørgenson

Table 2 Ranking of the discharge to the Dutch coastal zone (DCZ) of mandatory and recommended Oslo Paris (OSPAR) Commission[a] substances in 2002

Ranking	Substance	Sum of riverine and direct input to DCZ (t)
1[a]	Total N	429,000
2[a]	NO_3-N	309,000
3[a]	NH_4-N	16,000
4[a]	Total P	29,000
5[a]	PO_4-P	10,000
6[a]	Zn	2,159
7[a]	Cu	528
8[a]	Pb	386
9	Iopromide	22.6
10	Carbamazepine	12.1
11[a]	Cd	8.8–10
12	Diatrizoic acid	9.1
13	Iopamidol	8.8
14	Iomeprol	6.4
15	Sulfamethoxazol	5.3
16	Sotalol	4.8
17	Iohexol	4.7
18	Diclofenac	4.1
19[a]	Hg	3.3
20	Metoprolol	2.9
21[a]	PCBs	0.217
22[a]	γ-HCH (lindane)	0.041 0.055

[a] OSPAR (2004)
PCBs polychlorinated biphenyls

and Halling-Sørenson 2000; Sanderson et al. 2004). According to Sanderson et al. (2004), ecotoxicological data are available in the open-peer reviewed literature or ecotoxicological databases [ECETOX (EU) and ECOTOX (US)], for fewer than 1% of human pharmaceuticals.

Sanderson et al. (2004) developed a quantitative tool for ranking and prioritizing environmental risks of pharmaceuticals based on (1) their ranked predicted aquatic toxicity, (2) removal rates in STPs, (3) potential to bioaccumulate, and (4) number of compounds comprising each class (used as a surrogate for volumes because classes with large diversity typically have higher volumes than do classes with low diversity). According to this tool, additives to pharmaceuticals (e.g., paraffin, anionic, and nonionic surfactants) were the most toxic classes and are not taken into account in this study. However, these substances are not the AIs of pharmaceuticals. Using this prioritizing tool, cardiovascular, gastrointestinal, antiviral, anxiolytic sedatives, hypnotics and antipsychotics, corticosteroid, and thyroid pharmaceuticals were predicted to be the most hazardous therapeutic classes. Predictions for individual compounds, within classes or for the entire database, are available from Sanderson (hsander@uoguelph.ca).

Stockholm County Council (2005) carried out an environmental hazard assessment of individual pharmaceuticals, based on three properties: persistence (P), bioaccumulation (B), and toxicity (T). Each of these properties is assigned a value on a 0-3 scale. The sum of the values constitutes the Persistence/Bioaccumulation/Toxicity (PBT) index. A PBT index of 0 indicates that the substance is readily biodegradable, does not bioaccumulate, and has low ecotoxicity. A PBT index of 9 indicates that the substance is not readily biodegradable, has a potential for bioaccumulation, and has very high ecotoxicity. For example, the pharmaceuticals tamoxifen (antiestrogen) and ethinylestradiol (sex hormone) each have a PBT index of 9.

Only 5 of the top 20 pharmaceuticals discharged to the DCZ (Table 2) are classified by the Stockholm County Council (2005). The PBT indices for carbamazepine, sulfamethaxazol, diclofenac, metoprolol, and ibuprofen are 4, 3, 7, 4, and 5, respectively. Diclofenac has the highest PBT index. This pharmaceutical is very resistant (not readily biodegradable), bioaccumulates, and is moderately toxic: lethal, effective, and inhibitory concentrations (50% affected) ranged between 10 mg L^{-1} and 100 mg L^{-1}. Only three pharmaceuticals mentioned in this study have a PBT index of 9: propanolol, norfloxacin, and tamoxifen.

Unfortunately, the X-ray-contrast-media agents, which are most ubiquitously present in Dutch surface-, sewage-, and industrial water, are not classified by the Stockholm County Council (2005). However, Steger-Hartman et al. (1998, 1999) reported that environmental sublethal effects to organisms are unknown for X-ray contrast media. Moreover, no toxic effects were observed in *Daphnia magna*, using the chronic reproduction test, with concentrations up to 1 g L^{-1}, or in bacteria (*Vibrio fisheri, Pseudomonas putida*), algae (*Scenedesmus subspicatus*), crustaceans (*Daphnia magna*), and fish (*Danio rerio, Leuciscus*), using short-term acute ecotoxicological test systems containing up to 10 g L^{-1} of iopromide. Although it is unlikely that X-ray contrast media pose an immediate threat to the environment, their environmental presence at relatively high concentrations, high persistence, and unknown sublethal effects compel further environmental research and assessment.

4 Summary

This chapter assessed the annual median discharge of over 100 pharmaceuticals to the DCZ. Calculations were based on pharmaceutical concentrations in surface-, sewage-, and industrial water in the Netherlands.

In 2002, riverine discharge to the DCZ for individual pharmaceuticals varied from 0 (concentration below LOD) to 27 t yr^{-1}. However, in 2002 the annual amount was less than 2 t for 75% of the studied pharmaceuticals. The highest loads were calculated for X-ray contrast media, that is, values of 18–27 t yr^{-1} for iopromide.

The top 20 pharmaceuticals discharged by rivers to the DCZ are dominated by X-ray contrast media ($n = 7$), followed by antibiotics ($n = 6$), analgesics/antipyretics/anti-inflammatory drugs ($n = 2$), beta blockers ($n = 2$), fibrates/lipid regulators

($n = 1$), veterinary antibiotics ($n = 1$), and others ($n = 1$). The direct discharge (sewage water and industrial water) of pharmaceuticals to the DCZ, in 2002, for individual pharmaceuticals varied from <0.0009 to 0.27 t yr^{-1} for sewage water, and from <0.0009 to 0.33 t yr^{-1} for industrial wastewater. The highest loads were calculated for sotalol (beta blocker) and diatrizoic acid (X-ray contrast medium) in sewage water and industrial wastewater, respectively. The direct discharge of pharmaceuticals to the DCZ is <5% of that from riverine discharge.

The discharge of these pharmaceuticals to the DCZ in 2002 is in the same order of magnitude as the discharge rates of the mandatory OSPAR substances Cd (8.8–10 t yr^{-1}) and Hg (3.3 t yr^{-1}), and are higher than the discharge rates of the mandatory substance lindane (0.041-0.055 t yr^{-1}) and the recommended substance PCBs (0.217 t yr^{-1}).

Although some pharmaceuticals are discharged in significant amounts to the DCZ, the human and ecotoxicological risks of these highly biologically active compounds are largely unknown. To determine the environmental hazard and risk of discharged pharmaceuticals to the marine environment, future research should focus on a baseline study and a risk assessment of the discharged pharmaceuticals in the DCZ.

Acknowledgments The authors acknowledge Dr. M. Schrap and G. Rijs for kindly delivering the data and correcting the manuscript. H. van Reekum constructed the map.

References

Barreveld HL, Berbee RPM, Ferdinandy MMA (2001) 'Vergeten' stoffen in Nederlands oppervlaktewater. RIZA rapport 2001.020.

Belfroid AC, Van der Horst A., Vethaak AD, Schäfer AJ, Rijs GBJ, Wegener J, Cofino WP (1999) Analysis and occurrence of estrogenic hormones and their glucuronides in surface water and waste water in the Netherlands. Sci Total Environ 225:101-108.

Blom A ThG, de Bruijn JCMJ, de Jong JGAM (1995) Verspilling. Faculteit Pharmacie, Universiteit Utrecht, Utrecht, the Netherlands, 69(XXI).

Boxall ABA, Kolpin DW, Halling-Sørensen B, Tolls J (2003) Are veterinary medicines causing environmental risks? Environ Sci Technol 37:286A-294A.

Buser HR, Muller MD, Theobald N (1998a) Occurrence of the pharmaceutical drug clofibric acid and the herbicide mecoprop in various Swiss lakes and in the North Sea. Environ Sci Technol 32:188-192.

Buser HR, Poiger T, Muller MD (1998b) Occurrence and fate of the pharmaceutical drug diclofenac in surface waters: Rapid photodegradation in a lake. Environ Sci Technol 32:3449-3456.

Buser HR, Poiger T, Muller MD (1999) Occurrence and environmental behaviour of the chiral pharmaceutical drug ibuprofen in surface waters and in wastewater. Environ Sci Technol 33:2529-2535.

Carbella M, Omil F, Lema JM (2007) Calculations methods to perform mass balance of micropollutants in sewage treatment plants. Application to pharmaceutical and personal care products (PPCPs). Environ Sci Technol 41(3):884-890.

Comoretto L, Chiron S (2005) Comparing pharmaceutical and pesticide loads into a small Mediterranean river. Sci Total Environ 349:201-210.

Derksen JGM, van Eijnatten GM, Lahr J, van der Linde P, Kroon AGM (2001) Milieu-effecten van humane geneesmiddelen. RIZA rapport 2001.051.

Fent K, Weston AA, Caminada D (2006) Ecotoxicology of human pharmaceuticals. Aquat Toxicol 76(2):122-159.

Hamscher G, Pawelzick HT, Nau H, Hartung J (2002) Detection of antibiotics in dust originating from pig farms. In: Proceedings of the 21th SETAC Europe Meeting, Vienna, 2002. SETAC Europe, Brussels, Belgium, p. 11.

Hartig C, Storm T, Jekel M (1999) Detection and identification of sulphonamide drugs in municipal waste water by liquid chromatography coupled with electrospray ionisation tandem mass spectrometry. J Chromatog A 854:163-173.

Hartmann A, Alder AC, Koler T, Widmer RM (1998) Identification of fluoroquinolone antibiotics as the main source of umuC genotoxicity in native hospital wastewater. Environ Toxicol Chem 17:377-382.

Heberer T (2002) Occurrence, fate and removal of pharmaceutical residues in the aquatic environment: A review of recent data. Toxicol Lett 131:5-17.

Highnite C, Azarnoff DL (1977) Drugs and drug metabolites as environmental contaminants: Chlorophenoxyisobutyrate and salicylic acid in sewage water effluent. Life Sci 20:337-342.

Hirsch R, Ternes TA, Haberer K, Kratz K (1996) Determination of betablockers and sympathomimetrics in the aquatic environment. Vom Wasser 87:263-274.

Hirsch R, Ternes K, Haberer K, Kratz (1999) Occurence of antibiotics in the aquatic environment. Sci Total Environ 225:109-118.

Jørgenson SE, Halling-Sørenson B (2000) Editorial: Drugs in the environment. Chemosphere 40:691-699.

Klijnstra JW (2005) Korte bureaustudie naar het voorkomen van farmaceutica in antifoulingsverven. TNO-rapport CA05.8057.

Mons MN, van Genderen J, van Dijk-Looijaard AM (2000) Inventory of the presence of pharmaceuticals in Dutch water. KIWA rapport WR6, Nieuwegein, The Netherlands, 38 pp.

Mons MN, Hogenboom A, Noij THM (2003) Pharmaceuticals and drinking water supply in the Netherlands. KIWA report no. BTO 2003.040, Nieuwegein, The Netherlands, 76 pp.

Netherlands Hydrological Society (2004) Water in the Netherlands. Delft, the Netherlands, 186 pp.

Oslo-Paris (OSPAR) (2004) Overview of the results of the comprehensive study on riverine inputs and direct discharges (RID) from 2000 to 2002, ISBN 1-904426-58-1.

Peterson SM, Batley GE, Scammell MS (1993) Tetracycline in antifouling paints. Mar Pollut Bull 26(2):96-100.

Richardson ML, Bowron JM (1985) The fate of pharmaceutical chemicals in the aquatic environment. J Pharm Pharmacol 37:1-12.

Rijs GBJ, Laane RWPM, de Maagd G-J (2003) Voorkomen is beter dan genezen. Een beleidsanalyse over 'geneesmiddelen en watermilieu', RIZA rapport 2003.037, Lelystad, The Netherlands.

Rogers IH, Birtwell IK, Kruzynski GM (1986) Organic extractables in municipal wastewater, Vancouver, British Columbia. Water Pollut Res J Can 21:187-204.

Sacher F, en Stoks P (2003) Pharmaceutical residues in waters in the Netherlands. Results of a monitoring programme for RIWA. RIWA report, ISBN 90-6683-106-5, Nieuwegein, The Netherlands.

Sanderson H, Johnson DJ, Reitsma, T, Brain RA, Wilson CJ, Solomon KR (2004) Ranking and prioritization of environmental riks of pharmaceuticals in surface waters. Regul Toxicol Pharmacol 39:158-183.

Schrap S, Rijs GBJ, Beek MA, Maaskant JFN, Staeb J, Stroomberg G, Tiesnitsch J (2003) Humane en veterinaire geneesmiddelen in Nederlands oppervlaktewater en afvalwater. RIZA rapport 2003.023, Lelystad, The Netherlands.

Stan H-J, Heberer T (1996) Occurrence of polar organic contaminants in Berlin drinking water. Vom Wasser 86:19-31.

Stan H-J, Linkerhägner M (1992) Identifizierung von 2-(4-Chlorphenoxy)-2-methyl-propionsäure im Grundwasser mittels Kapillar-Gaschromatographie mot Atomemissionsdetektion und Massenspektrometrie. Vom Wasser 79:75-88.

Stan H-J, Heberer T, Linkerhägner M (1994) Occurrence of clofibric acid in the aquatic system – Is the use in human medical care the source of the contamination of surface, ground, and drinking water? Vom Wasser 83:57-68.

Steger-Hartmann Th, Länge K, Schweinfurth H (1998) Umweltverhalten und ökotoxikologische Bewertung. von iodhaltigen Röntgenkontrastmitteln. Vom Wasser 91:185-194.

Steger-Hartmann Th, Länge K, Schweinfurth H (1999) Environmental risk assessment for the widely used iodinaded x-ray contrast agent iopromide (Ultravist). Ecotoxicol Environ Saf 42(3):274-281.

Stockholm County Council (2005) Environmentally classified pharmaceuticals (www.janusinfo.se).

Stumpf M, Ternes TA, Haberer K, Seel P, Baumann W (1996) Sci Tot Environ 225:135-141.

Ternes A (1998) Occurrence of drugs in German sewage treatment plants and rivers. Water Res 32(11):3245-3260.

Ternes TA, Hirsch R (2000) Occurrence and behaviour of X-ray contrast media in sewage facilities and the aquatic environment. Environ Sci Technol 34:2741-2748.

Tixier C, Singer H, Oellers S, Müller S (2003) Occurrence and fate of carbamapezine, clofibric acid, diclofenac, ibuprofen, ketoprofen and naproxen in surface waters. Environ Sci Technol 37(6):1061-1068.

Versteegh JFM, Stolker AAM Niesing W, Muller JJA (2003) Geneesmiddelen in drinkwater en drinkwaterbronnen. Resultaten van meetprogramma 2002. RIVM rapport 703719004/2003, Bilthoven, The Netherlands.

Vethaak AD, Rijs GBJ, Schrap SM, Ruiter H, Gerritsen A, Lahr J (2002) Estrogens and xeno-estrogens in the aquatic environment of The Netherlands. Occurrence, potency, and biological effects. RIZA/RIKZ report no. 2002.001, Lelystad, The Netherlands.

Waterbase (2006) www.waterbase.nl

Weigel S (2003) Occurrence, distribution and fate of pharmaceuticals and further polar contaminants in the marine environment. Universität Hamburg.

Weigel S, Berger U, Jensen E, Kallenborn R, Thoresen H, Hühnerfuss H (2004) Determination of selected pharmaceuticals and caffeine in sewage and seawater from Tromso/Norway with emphasis on ibuprofen and its metabolites. Chemosphere 56:583-592.

Weil H, Knepper TP (2006) Pharmaceuticals in the River Rhine. In: The Rhine, Knepper TP (ed.) The Handbook of Environmental Chemistry Vol. 5: Water Pollution Part C, Springer Verlag, Berlin, pp. 177-184.

Chapter 2
Methods for Deriving Pesticide Aquatic Life Criteria

Patti L. TenBrook, Ronald S. Tjeerdema, Paul Hann, and Joseph Karkoski

Contents

P.L. TenBrook
Current affiliation: U.S. EPA Region 9, 75 Hawthorne Street, San Francisco, CA 94105.

R.S. Tjeerdema (✉)
Department of Environmental Toxicology, College of Agricultural and Environmental Sciences, University of California, Davis, CA 95616-8588
E-mail: rstjeerdema@ucdavis.edu

P. Hann, J. Karkoski
Central Valley Regional Water Quality Control Board, 11020 Sun Center Drive #200, Rancho Cordova, CA 95670.

D.M. Whitacre (ed.) *Reviews of Environmental Contamination and Toxicology* Volume 199,
doi: 10.1007/978-0-387-09808-1_2,

1 Introduction

The mission of California's nine Regional Water Quality Control Boards (RWQCB) is "to develop and enforce water quality objectives and implement plans which will best protect the beneficial uses of the State's waters, recognizing local differences in climate, topography, geology and hydrology" (California SWRCB 2005). To accomplish that mission, each RWQCB is responsible for development of a "basin plan" for its hydrologic area. The "Water Quality Control Plan (Basin Plan) for the Sacramento River and San Joaquin River Basins" (CVRWQCB 2004) contains the following excerpts regarding toxic substances in general, and pesticides in particular:

> ...waters shall be maintained free of toxic substances in concentrations that produce detrimental physiological responses in human, plant, animal, or aquatic life.
>
> No individual pesticide or combinations of pesticides shall be present in concentrations that adversely affect beneficial uses.
>
> Discharges shall not result in pesticide concentrations in bottom sediments or aquatic life that adversely affect beneficial uses.
>
> Pesticide concentrations shall not exceed the lowest levels technically and economically achievable.

Development of specific numeric criteria for more pesticides would provide clear goals for permitting and Total Maximum Daily Load (TMDL) programs. This chapter is part of a larger project to develop a new aquatic life criteria methodology for pesticides. The intent of this methodology is to establish, from available toxicity data, a concentration unlikely to produce detrimental physiological effects in aquatic species. The Central Valley RWQCB requested a literature review to define the following: (1) criteria derivation methodologies currently in use, or proposed for use, throughout the world; (2) original studies supporting the methodologies; (3) proposed modifications of existing methodologies; and (4) relevant and recent research in ecotoxicology and risk assessment. Four documents were found that provide a good overview of the latest scientific thinking in the field of water quality criteria derivation. First, is a book, *Reevaluation of the State of the Science for Water-Quality Criteria Development* (Reilly et al. 2003), which is a report of conclusions reached by participants in a Society of Environmental Toxicology and Chemistry (SETAC) Pellston workshop. Second, is the "Draft Report on Summary of Proposed Revisions to the Aquatic Life Criteria Guidelines" (USEPA 2002a). Third, is a report from the United Kingdom (UK) Environment Agency called "Derivation and Expression of Water Quality Standards, Opportunities and Constraints in Adopting Risk-Based Approaches in EQS Setting" (EQS: environmental quality standard; Whitehouse et al. 2004). The final document is a report from the Fraunhofer-Institute of Molecular Biology and Applied Ecology, prepared on behalf of the European Commission (EC), called "Towards the Derivation of Quality Standards for Priority Substances in the Context of the Water Framework Directive" (Lepper 2002). The information in these four reports, as well as conversations with state and federal regulators (Karkoski 2005, personal communication; Denton 2005, personal communication), were used to construct Table 1, which is a list of components to consider when evaluating and developing aquatic life water quality criteria derivation methodology.

Table 1 Components to be addressed by water quality criteria derivation methodology

Category	Component (listed alphabetically)	Reference
Criteria types/uses (Section 4)	One type/level of criterion versus multiple types/levels	a,c,d,e
	Use in regulatory programs	a,b,c,d
Protection level (Section 5)	Economically, ecologically, recreationally important species	f
	Ecosystem function and structure	a,c,d
	Individuals versus populations	c
	Justification of percentile levels (i.e., 10th, 5th, 1st)	b,c,d
	Probability of over- or underprotection	a,b,c,d
Ecotoxicity and physical-chemical data (Section 6)	Data quality and quantity	
	Acceptability criteria	b,c,d
	Minimum data set	a,b,c,d
	Minimum literature search	f
	Taxa number and diversity	a,b,c
	Ecological relevance	c,d
	Kinds of data	
	Acute (LC_x/EC_x, NOEC) versus chronic (EC_x, NOEC)	a,b,c,d
	Ecosystem, field, semi-field, laboratory	a,b,c,d
	Multispecies versus single-species	a,d
	Traditional versus nontraditional endpoints	a,b,c,d
Criteria calculation (Section 7)	Bioaccumulation/secondary poisoning	a,b,d
	Community/ecosystem/population/foodweb models	a,c,d
	Confidence limits for criteria/explicitly stated uncertainty	a,b,c,d
	Degree of aggregation of taxa	b
	Derivation and justification of assessment/uncertainty factors	b,c,d,e
	Encouragement data generation	c,d
	Environmental fate of chemicals	a,d
	Exposure considerations	
	Bioavailability	a,d
	Short-term/acute (including pulse) and long-term/chronic	a,b,c,d
	Magnitude, duration and frequency	a,b,c,d
	Monitoring considerations	d
	Recovery from toxic events	a,b
	Harmonization/coherence across media	a,d
	Incorporation of physical–chemical data	a,b,d
	Kinetic-based modeling/time to event analysis	a,b,c
	Mixtures/multiple stressors	a,b,d
	Multipathway (e.g., dietary) exposure	a,b
	Plants and animals combined versus separate	a,b,d
	Risk assessment approach	a,b,c,d
	Separate acute and chronic criteria versus single criterion	a,b
	Site specificity	a,b,c
	Small data sets	a,b,e

(continued)

Table 1 (continued)

Category	Component (listed alphabetically)	Reference
	Species sensitivity distributions (SSD)	a,b,c,d
	Toxicant mode of action	a,d
	Threatened and endangered species	a,b
	Wildlife	a,d
	Utilization of available data	c,d,e

"Section" of this chapter in which the specified components are addressed
[a]Reiley et al. 2003
[b]USEPA 2002a
[c]Whitehouse et al. 2004
[d]Lepper 2002
[e]Personal communication (Karkoski 2005; Denton 2005)
[f]Not part of discussions in references a–d but are part of existing criteria derivation methodologies in the Unites States (USEPA 1985), Australia/New Zealand (ANZECC and ARMCANZ 2000), and/or the Netherlands (RIVM 2001)

In this chapter, the components listed in Table 1 are discussed with respect to how they are, or are not, addressed by existing criteria derivation methodologies. Included in the discussion are methodologies from (listed alphabetically) Australia/New Zealand, Canada, Denmark, the European Union/European Commission (EU/EC), France, Germany, the Netherlands, the Organization for Economic Co-operation and Development (OECD), South Africa, Spain, the United Kingdom (UK), and the United States (US), including the Great Lakes Region, and a few individual US states whose methodologies diverge somewhat from USEPA guidance (1985). In some cases, original documents were not available in English, but other resources containing summaries of those documents were available and were used for this chapter. Existing methodologies are evaluated against recent research and reports on criteria derivation techniques. In addition to pesticides, most of the methodologies address toxicity from metals and other inorganic chemicals (e.g., ammonia), and nonpesticide organic chemicals. This chapter is focused on methodologies to derive pesticide criteria for the protection of aquatic life. Some of the latest recommendations for water quality criteria derivation methodologies are simply not technically feasible, at this time, because of a paucity of data or lack of agreement among experts on techniques.

Table 2 is provided as a reference to help the reader with the many acronyms used throughout this chapter.

2 Summary of Major Methodologies Reviewed

Many existing methodologies are discussed in this chapter, but the focus is on a few that are widely accepted and used (USEPA 1985; RIVM 2001 – updated from VROM – Ministry of Housing, Spatial Planning and Environment, The Hague, the Netherlands), represent unique approaches (CCME Council of Ministers of the Environment – 1999), or constitute newer methodologies that incorporate and

Table 2 List of acronyms and abbreviations used in this chapter

AA	Annual average
ACE	Acute-to-chronic Estimation
ACR	Acute-to-chronic Ratio
AEV	Acute effect value
AF	Assessment factor
ANZECC	Australia and New Zealand Environment and Conservation Council
ARMCANZ	Agriculture and Resource Management Council of Australia and New Zealand
ASTM	American Society for Testing and Materials
BAF	Bioaccumulation factor
BC	British Columbia
BCF	Bioconcentration factor
CAS	Chemical Abstract Service
CCC	Criterion continuous concentration
CCME	Canadian Council of Ministers of the Environment
CDFG	California Department of Fish and Game
CEV	Chronic effect value
CMC	Criterion maximum concentration
CVRWQCB	Central Valley Regional Water Quality Control Board
CSTE/EEC	Scientific Advisory Committee on Toxicity and Ecotoxicity of Chemicals/European Economic Community
CWA	Clean Water Act
DTA	Direct toxicity assessment
ECB	European Chemicals Bureau
EC	European Commission
EC_x	Concentration that affects x% of exposed organisms
ECL	Environmental concern level
ECOTOC	European Centre for Ecotoxicology and Toxicology of Chemicals
ERL	Environmental risk level
EINECS	European Inventory of Existing Commercial Substances
EqP	Equilibrium partitioning
EQS	Environmental quality standard
EU	European Union
FACR	Final acuteto-chronic ratio
FAV	Final acute value
FCV	Final chronic value
FPV	Final plant value
FRV	Final residue value
GESAMP	Group of Experts on the Scientific Aspects of Marine Protection
GMAV	Genus mean acute value
HC_x	Hazardous concentration potentially harmful to x% of species
ICE	Interspecies correlation estimation
IUPAC	International Union of Pure and Applied Chemistry
KemI	Swedish Chemicals Agency
K_H	Henry's law constant
K_{ow}	Octanol–water partition coefficient
Kp	Solid–water partition coefficient
LC_x	Concentration lethal to x% of exposed organisms

(continued)

Table 2 (continued)

LOEC	Lowest observed effect concentration
LOEL	Lowest observed effect level
MAC	Maximum allowable concentration
MATC	Maximum acceptable toxicant concentration
MITI	Ministry of International Trade and Industry, Japan
MPC	Maximum permissible concentration
msPAF	Multispecies potentially affected fraction
MTC	Maximum tolerable concentration
NC	Negligible concentration
NCDENR	North Carolina Department of Environment and Natural Resources
NOEC	No-observed effect concentration
OECD	Organization for Economic Co-operation and Development
parNEC	Parametric no-effect concentration
pK_a	Acid dissociation constant
PNEC	Probable no-effect concentration
QSAR	Quantitative structure activity relationship
QSSR	Quantitative species sensitivity relationship
QT	Quality target
RI	Reliability index
RIVM	National Institute of Public Health and the Environment, Bilthoven, the Netherlands
RPF	Relative potency factor
RWQCB	Regional Water Quality Control Board
SACR	Secondary acute-to-chronic ratio
SAV	Secondary acute value
SCC	Secondary chronic concentration
SCV	Secondary chronic value
SETAC	Society of Environmental Toxicology and Chemistry
SMC	Secondary maximum concentration
SMCV	Species mean chronic value
SMAV	Species mean acute value
SRC_{ECO}	Ecosystem serious risk concentration
SSD	Species sensitivity distribution
SWRCB	State Water Resources Control Board
TES	Threatened and endangered species
TEF	Toxic equivalency factor
TGD	European Union's Technical Guidance Document on Risk Assessment
TMDL	Total maximum daily load
TRG	Tissue residue guideline
TSD	Technical Support Document for Water Quality-based Toxics Control
TV	Trigger value
UK	United Kingdom
US	United States
USEPA	United States Environmental Protection Agency
VROM	Ministry of Housing, Spatial Planning and Environment, The Hague, The Netherlands
WCS	Water-based Criteria Subcommittee
WER	Water effect ratio
WFD	Water framework directive

improve upon the best features of prior methodologies (ANZECC and ARMCANZ 2000; USEPA 2003a). The European Union's Technical Guidance Document (TGD) on Risk Assessment is also emphasized because it represents the latest European guidance on derivation of predicted no-effect concentrations (PNECs) for risk assessments, and EU member nations are starting to use the TGD for derivation of water quality criteria (Traas, 2005, personal communication).

Table 3 lists the six major methodologies commonly employed, the types of criteria that are derived from them, and how the criteria are used. The USEPA methodology (1985) utilizes a statistical extrapolation procedure to derive criteria, while the Canadian methodology (CCME 1999) utilizes an assessment factor (AF) approach. All of the others utilize a combination of these two basic criteria derivation methods. Similarities and differences between key elements of the methodologies identified in Table 3 are summarized in Table 4. As this chapter proceeds, Table 4 should help the reader understand how each element fits into existing criteria derivation methodologies.

3 Water Quality Policy

Different countries of the world have different environmental policies, which are reflected in their water quality criteria derivation methodologies. The EU's Water Framework Directive (WFD) is the policy document guiding water quality protection efforts for EU member states (EU 2000). The 11th principle of the WFD states that environmental policy should "...contribute to pursuit of the objectives of preserving, protecting and improving the quality of the environment, in prudent and rational utilization of natural resources, and to be based on the precautionary principle and on the principles that preventive action should be taken, environmental damage should, as a priority, be rectified at source and that the polluter should pay." The precautionary principle may be summed up as follows:

> In order to protect the environment, the precautionary approach shall be widely applied by States according to their capabilities. Where there are threats of serious or irreversible damage, lack of full scientific certainty shall not be used as a reason for postponing cost-effective measures to prevent environmental degradation. (Rio Convention 1992)

Applegate (2000) affirms that, while containing many precautionary elements, US policy does not adhere to the precautionary principle; many other factors (especially economics) drive US environmental policy. In addition, the USEPA has embraced the use of ecological risk assessment to assess potential chemical hazards. Chapman et al. (1998) noted that the precautionary principle undermines the risk assessment approach by defining either infinitely small no-effect concentrations or infinitely large safety factors. Although subscribing to the precautionary principle, EU member countries, Canada, the Netherlands, South Africa, Denmark, and Australia/New Zealand, have incorporated risk assessment techniques into their water quality criteria derivation methodologies (Lepper 2002; European Chemical Bureau, ECB 2003; CCME 1999; RIVM 2001; Roux et al. 1996; Samsoe-Petersen and Pedersen 1995; ANZECC and ARMCANZ 2000). Thus, although arising from

Table 3 Overview of major derivation methodologies and description of their criteria

Method title	Source	Yr	Country	Criterion	Criterion description
Guidelines for deriving numerical national water quality criteria for the protection of aquatic organisms and their uses	USEPA	1985	United States	CMC: criterion maximum concentration	Used for setting water quality standards, setting discharge limits, and other regulatory programs; for protection from short-term exposure
				CCC: criterion continuous concentration	Used for setting water quality standards, setting discharge limits, and other regulatory programs; for protection from long-term exposure
A protocol for the derivation of water quality guidelines for the protection of aquatic life	CCME	1999	Canada	Guidelines	Single maximum which is not to be exceeded
Australia and New Zealand guidelines for fresh and marine water quality	ANZECC & ARMCANZ	2000	Australia/ New Zealand	HRTV: high reliability trigger value	Derived from >1 multispecies or >5 single-species chronic values; not a mandatory standard; exceedance triggers further investigation
				MRTV: medium reliability trigger value	Derived from >5 acute data; not a mandatory standard; exceedance triggers further investigation
				LRTV: low reliability trigger value	Derived from <5 acute or chronic values; not used as a guideline value
Guidance document on deriving environmental risk limits in the Netherlands	RIVM	2001	The Netherlands	NC: negligible concentration	Used to set environmental quality standards (EQS); EQS may or may not be legally binding
				MPC: maximum permissible concentration	Used to set EQS; EQS may or may not be legally binding
				SRC_{ECO}: ecosystem serious risk concentration	Used to set EQS; EQS may or may not be legally binding
Water quality guidance for the Great Lakes system	USEPA	2003	United States	Tier I CMC	Adopted into water quality standards or used to implement narrative criteria; for protection from short-term exposure
				Tier I CCC	Adopted into water quality standards or used to implement narrative criteria; for protection from long-term exposure
				Tier II CMC	Used only for implementation of narrative criteria; for protection from short-term exposure
				Tier II CCC	Used only for implementation of narrative criteria; for protection from long-term exposure
Technical guidance document on risk assessment, Part II. Environmental risk assessment.	ECB	2003	European Union	PNEC: predicted no-effect concentration	Used in risk assessment

Table 4 Summary of similarities and differences between key elements of six major criteria derivation methodologies; ✓ = element of methodology; blank cell = not an element of methodology

		Data used directly for derivation					SSD[1] method							AF[2] method		Criteria components									
Country Region	Reference	Sources	Evaluation Criteria	QSARS[3] allowed	Multiecies data	Endpoints linked to SGR[4]	Log-triangular	Log normal	Burr family/best fit	Minimum number of values required	Minimumnumber of taxa required	Uncertainty quantified	All data used	Minimum number of values required	Minimum number of taxa required	Acute	Chronic	Magnitude	Duration	Frequency	Bioaccumulation	Mixtures	Bioavailability	Water quality	TES[5]
United States	USEPA (1985)		✓			R[6]	✓			8				6-9	5	✓	✓	✓	✓	✓	✓		✓	✓	
Canada	CCME (1999)		✓			S[7]								1	1		✓	✓							
Australia/New Zealand	ANZEC/ARMCANZ (2000)		✓	✓	✓				✓	5	5	✓	✓	1	1		✓	✓			✓		✓	✓	
The Netherlands	RIVM (2001)	✓	✓	✓				✓		4	4	✓	✓	1	1		✓	✓			✓				✓
United States/Great Lakes	USEPA (2003)		✓				✓		✓	8	8			1	1	✓	✓	✓	✓	✓			✓	✓	✓
European Union	ECH (2003)		✓		✓					10	8	✓	✓	1	1		✓	✓			✓				

[1]Species sensitivity distribution
[2]Assessment factor
[3]Quantitative structure activity Relationship
[4]Survival/growth/reproduction
[5]Threatened and endangered Species
[6]Rarely
[7]Secondary data only

different policy tenets, many of the water quality criteria derivation techniques used throughout the world are applicable under US and California policy.

4 Criteria Types and Uses

Three types of water quality criteria are described by the USEPA: numeric, narrative, and operational (USEPA 1985). This chapter is concerned with derivation of numeric criteria, which can be used in setting water quality standards. This section describes many different types of numeric criteria that may be derived according to various methodologies, depending on how the values are to be used and how much data are available.

In the literature, numeric water quality criteria are referred to by many different terms. For example, they may be called trigger values (TVs; ANZECC and ARMCANZ 2000), guidelines (CCME 1999), criteria (USEPA 1985; Samsoe-Petersen and Pedersen 1995; Roux et al. 1996), quality standards, thresholds (Lepper 2002; Zabel and Cole 1999), environmental risk limits (ERLs; RIVM 2001), maximum tolerable concentrations (MTCs; OECD 1995), predicted no-effect concentrations (PNECs; ECB 2003), water quality objectives (WQOs; Bro-Rasmussen et al. 1994), and quality targets (BMU 2001; Irmer 1995). The common thread in all of these is that the values derived are scientifically based numbers which are intended to protect aquatic life from adverse effects of pesticides, without consideration of defined water body uses, societal values, economics, technical feasibility, or other nonscientific considerations. This definition corresponds to what the USEPA calls a numeric criterion (as per Section 304(a) of the Clean Water Act), and it is the derivation of this type of number that is the subject of this chapter.

4.1 *Numeric Criteria Versus Advisory Concentrations*

In the US, numeric criteria are derived for compounds when adequate toxicity, bioaccumulation, and/or field data are available (USEPA 1986). These criteria may be used for such things as developing water quality standards or setting effluent limitations (USEPA 1985). If adequate data are not available for criteria derivation, advisory concentrations are then derived. Advisory concentrations are used to interpret ambient water quality data. For example, if the ambient concentration of a chemical is below the advisory concentration, then there is no further concern; if the concentration is above the advisory concentration, then more data are collected, preferably enough to allow calculation of a criterion (USEPA 1986).

The USEPA Great Lakes water quality guidance (USEPA 2003a) provides for derivation of Tier I and Tier II criteria. Tier I criteria, which are derived from complete data sets according to the USEPA methodology (1985), may be adopted as numeric criteria, may be used to adopt water quality standards, or may be used to implement narrative criteria. Tier II criteria, similar to USEPA advisory concentrations, are derived from incomplete data sets using methodology similar to those of USEPA (1986), and are used only for implementation of narrative criteria.

4.2 Numeric Criteria of Different Types and Levels

Many existing criteria derivation methodologies include procedures for derivation of more than one level or type of criterion for each toxicant (OECD 1995; ANZECC and ARMCANZ 2000; La Point et al. 2003; RIVM 2001; Lepper 2002; USEPA 2003a). This refers either to (1) the derivation of different levels of criteria to meet different regulatory goals, or (2) to the use of ecological risk assessment techniques with increasing levels of technical sophistication, leading to criteria with site-specific application and greater certainty (La Point et al. 2003). The second of these is directly related to how much and what kinds of data are available for criteria derivation. Note that derivation of separate acute and chronic criteria, as is done in the USEPA (1985) and UK methodologies (Zabel and Cole 1999), is not what is meant here by "different types and levels" of criteria.

Compartment-specific ERLs are derived in the Netherlands (RIVM 2001). The three levels of ERLs are the ecosystem serious risk concentration (SRC_{ECO}), the maximum permissible concentration (MPC), and the negligible concentration (NC). The NC (concentration causing negligible effects to ecosystems) is calculated as the MPC divided by a safety factor of 100, and represents a regulatory target value. The MPC is a concentration that should protect all species in ecosystems from adverse effects. If concentrations in ambient waters are above the MPC, discharges can be further regulated. Between the MPC and the NC, principles of ALARA (as low as reasonably achievable) are used to continue reducing levels toward the NC (Warmer and Van Dokkum 2002). The SRC_{ECO} is a concentration at which ecosystem functions will be seriously affected, or are threatened to be negatively affected (assumed to be when 50% of species and/or 50% of microbial and enzymatic processes are potentially affected; RIVM 2001). Waters exceed the SRC_{ECO} require cleanup intervention efforts.

In the French methodology (Lepper 2002), four threshold levels, corresponding to biological quality and suitability classes for water bodies, are calculated for each substance. Threshold level 1 indicates negligible risk for all species, and is derived from either chronic or acute toxicity data, with safety factors applied. The level 2 threshold indicates possible risk of adverse effects for the most sensitive species, and is derived from the same data as level 1, but smaller safety factors are applied. Levels 3 and 4 indicate probable or significant risk of adverse ecosystem effects, respectively, and are derived solely from acute toxicity data. Tentative standards may be set when a minimum data set is not available. Freshwater standards may be used as tentative marine standards if insufficient marine data are available and there is no reason to suspect greater sensitivity among marine species. None of the threshold values derived by the French methodology are enforceable; the values serve as references for assessments and development of action plans.

The OECD methodology (1995) recognizes three levels of aquatic effects assessment and derives MTCs for each level. An initial, or primary, assessment is based on laboratory toxicity data from only one or two representatives of primary producers, primary consumers and predators. An intermediate or refined assessment is based on results of chronic or semi-chronic laboratory tests. Field or semi-field studies are used for comprehensive assessments. MTCs derived by the OECD methodology (1995) are used to set environmental quality objectives. However, MTCs have different levels of reliability, depending on how they are derived. An MTC derived from quantitative structure activity relationships (QSARs) has lower status than one derived from acute toxicity tests; one derived from acute toxicity tests has lower status than one derived from chronic tests; an MTC derived from a reliable, representative field test has the highest status. Lower status MTCs are used for setting priorities, rather than for setting enforceable objectives.

In Australia and New Zealand, TVs of low, medium, and high reliability are derived (ANZECC and ARMCANZ 2000). The reliability rating is dependent on how much data support the value. Only medium and high reliability values are used as final guideline TVs. Low reliability values, which are similar to USEPA advisory concentrations, are interim figures, which, if exceeded, indicate the need for further data collection. High and medium reliability TVs are not pass/fail levels. If exceeded, a TV is reevaluated and refined in a site-specific assessment. Further regulatory action usually occurs only if the site-specific value is exceeded (although risk managers have the option of using the more conservative, national TVs as enforceable values).

By whatever name, all of the values discussed (including those not currently used in setting water quality standards or objectives) represent efforts to estimate concentrations of chemicals below which beneficial uses can reliably be regarded as protected. When data are limited, numeric criteria of low site-specificity and high uncertainty can be derived. As more data become available, criteria can be refined for better site-specificity and greater certainty (Di Toro 2003; La Point et al. 2003).

5 Protection and Confidence

Aquatic life water quality criteria are intended to protect aquatic life from exposure to toxic substances. But what really is the goal? Is it overall ecosystem protection, or protection of each individual in the ecosystem? And, how certain can one be of meeting that goal? This section discusses how aquatic life protection goals are stated in various derivation methodologies, and how those goals have to be approached, considering the need to extrapolate ecosystem effects from single-species toxicity data. It is also important to know that, with a quantified level of certainty, criteria are achieving the intended level of protection.

5.1 Levels of Biological Organization to Protect

Officials must decide what level of biological organization (defined in Table 5) their water quality criteria are intended to protect. Several derivation methodologies seek to protect individuals or species, expecting that by doing so, they will protect ecosystems. Canada's guiding principles for the development of freshwater aquatic life guidelines state that guidelines will consider all components of the ecosystem, and will be "set at such values as to protect all forms of aquatic life and all aspects of the aquatic life cycle" (CCME 1999). Similarly, the UK derives aquatic life EQSs for the protection of all aquatic species. The Netherlands has the goal of protecting all species in ecosystems from adverse effects (RIVM 2001).

Most of the reviewed methodologies specifically seek to protect aquatic ecosystems. Water quality criteria in South Africa "allow for the sustainable functioning of healthy and balanced aquatic ecosystems." This is achieved by developing criteria that are protective of representative key species from a variety of trophic groups (Roux et al. 1996). France derives threshold levels that will maintain water's suitability to support its biological function and other uses (Lepper 2002). The USEPA criteria are intended to protect "aquatic organisms and their uses," without specifically aiming to protect ecosystems. However, the methodology states that ecosystems can tolerate some stress and it is not necessary to protect all species at all times (USEPA 1985). Arguing that this feature of the USEPA methodology (1985)

Table 5 Definitions for levels of biological organization

Level	Definition	Reference
Individual	A single organism	*Webster's New Collegiate Dictionary* (1976)
Species	A taxonomic grouping of morphologically similar individuals who actually or potentially interbreed	Curtis and Barnes (1981)
Population	A group of individuals of one species that occupy a given area at the same time	Curtis and Barnes (1981)
Community	All of the organisms inhabiting a common environment and interacting with one another	Curtis and Barnes (1981)
Ecosystem	All organisms in a community plus the associated abiotic environmental factors with which they interact	Curtis and Barnes (1981)
Ecosystem structure	The spatial and temporal relationships of biotic and abiotic components that support energy flow and biogeochemical processes in an ecosystem	Curtis and Barnes (1981)
Ecosystem function	The processes by which energy flows and materials are cycled through an ecosystem	Curtis and Barnes (1981)
Ecosystem engineer	Species that directly or indirectly modulates the availability of resources to other species	Lawton (1994)
Keystone species	Species whose removal from a community would precipitate a further reduction in species diversity or produce other significant changes in community structure and dynamics	Daily et al. (1993)

did not meet the needs of California regulators, Lillebo et al. (1988) developed a criteria derivation methodology, specifically for use in California, designed to ensure full protection of aquatic biological resources. In Australia and New Zealand, the goal is "to maintain and enhance the 'ecological integrity' of freshwater and marine ecosystems, including biological diversity, relative abundance and ecological processes" (ANZECC and ARMCANZ 2000). German quality targets are designed to "maintain or restore a self-reproducing and self-regulating biocenosis of plants, animals, and microorganisms that is typical of the location concerned and is as natural as possible" (Irmer et al. 1995). The OECD guidelines provide methods for derivation of criteria "where no adverse effects on the aquatic ecosystem are expected" (OECD 1995). Denmark derives water quality criteria that are defined as ecotoxicological no-effect concentrations (Samsoe-Petersen and Pedersen 1995). The PNECs derived by the EU risk assessment methodology (ECB 2003), are intended to ensure "overall environmental protection," whereas, the Scientific Advisory Committee on Toxicity and Ecotoxicity of Chemicals/European Economic Community (CSTE/EEC) states that WQOs should permit all stages in the life of aquatic organisms to be successfully completed, should not produce conditions that cause organisms to avoid habitat where they would normally be present, should not result in bioaccumulation, and should not alter ecosystem function (Bro-Rasmussen et al. 1994; originally in CSTE/EEC 1987). The state of North Carolina seeks to ensure aquatic life propagation and maintenance of biological integrity (North Carolina Department of Environment and Natural Resources 2003). As discussed previously, the mandate of the Central Valley RWQCB is to maintain waters free of "toxic substances at concentrations that produce detrimental physiological responses in human, plant, animal, or aquatic life" (CVRWQCB 2004).

5.2 Portion of Species to Protect

Despite somewhat differing goals, all methodologies are forced to rely a great deal on single-species toxicity data to derive criteria. As pointed out in ECB (2003), two important assumptions are critical to these methodologies, which seek ecosystem protection by extrapolation from single-species laboratory ecotoxicity tests: (1) ecosystem sensitivity depends on the most sensitive species; and (2) protecting ecosystem structure protects community function. This approach is common throughout the world and results from the relative availability of data from single-species toxicity tests, compared to multispecies or ecosystem data.

A corollary assumption is that ecosystems can sustain some level of damage (e.g., to individuals or populations) from toxicants or other stressors, and subsequently recover with no lasting harm. This assumption is not completely supported in the literature. As discussed by Spromberg and Birge (2005a, b), whether or not population-level effect results from toxicity-induced physiological responses of individuals depends very much on life-history characteristics of the particular species.

Zabel and Cole (1999) point out that, in the case of algae, if a sensitive species were eliminated from an ecosystem, the photosynthetic function could be quickly replaced by another, less sensitive species. Ecosystem structure will have changed, but function is maintained. In contrast, Daily et al. (1993) note that the disappearance of a single species could lead to the unraveling of community structure as a result of complex interactions among species. Lawton (1994) explores the importance of "ecosystem engineers" and states that loss of keystone species, whether they are engineers or important trophic links, may cause dramatic and sudden ecosystem changes.

It would seem, then, that ecosystems may not be protected if water quality criteria are derived by a method that does not offer protection to 100% of species. However, there is no way to ensure such complete protection, because it is not possible to know the entire composition of an ecosystem. Even if an ecosystem were fully comprehended, it would not be possible to determine the sensitivity of all component species of that ecosystem. This chapter presents and evaluates alternative methods for estimating ecosystem no-effect concentrations (levels not detrimental to aquatic organisms) by extrapolating from available toxicity data, the bulk of which comprises single-species laboratory studies. To determine if numbers derived from these methods are adequately protective (i.e., meet policy goals), they must be validated in field or semi-field studies.

5.3 Probability of Over- or Underprotection

It is very useful to express criteria with confidence limits, because it provides environmental managers with insight on how likely it is that a criterion will provide the intended level of protection. Criteria that overprotect may result in unnecessary expenditures, whereas criteria that under protect may lead to ecosystem damage. Many criteria methodologies (Canada, France, Germany, and UK) involve compilation of data and then selection of the single most sensitive datum (often multiplied by an extrapolation factor) to represent the criterion (CCME 1999; Lepper 2002; Zabel and Cole 1999). Criteria so derived do not have confidence limits associated with them. Although the value set may be protective, there is no way to know to what degree the value over- or underprotects. Criteria derived by the USEPA methodology (1985) also do not have associated confidence limits, despite the fact that they use a species sensitivity distribution (SSD) methodology. Australia and New Zealand (ANZECC and ARMCANZ 2000), The Netherlands (RIVM 2001), and OECD (1995) use SSD techniques that derive criteria at specified confidence levels. For example, for a criterion derived at a 50% confidence level, the true no-effect level may be either above or below the derived criterion with equal probability. If derived at a 95% confidence level, there is only a 5% chance that the true no-effect level rests below the derived criterion. This kind of information can provide environmental managers with some sense as to the reliability of criteria.

6 Ecotoxicity and Physical–Chemical Data

At the core of all criteria derivation methodologies lies ecotoxicological effects data. Good criteria must be based on good quality data of adequate taxonomic diversity. Physical–chemical data are important for proper interpretation of toxicity test data, for estimation of bioavailability, and for estimation of toxicity for some classes of chemicals. Thus, criteria derivation methodologies must include clear guidance regarding how much of what kinds of data are required for calculation of criteria. A big challenge is finding ways to derive criteria from very small data sets. Ideally, it would be possible to derive scientifically sound criteria, from the minimum data sets typically required for pesticide registration procedures. The focus of this section is to review and address the quality and quantity of data required by existing methodologies.

6.1 Data Sources and Literature Search

Whatever the derivation methodology, the most reliable and most certain criteria are derived from the largest and best quality data sets. It is very helpful for a criteria derivation methodology to include some guidance on where and how to find data. To avoid perceptions, for example, that a regulator has selected only data from very sensitive species, or that a chemical producer has selected only data from very tolerant species, there should be explicit guidance regarding what constitutes a minimal literature search.

Of reviewed methodologies, the Dutch methodology provides the most detailed information regarding sources of ecotoxicological and physical–chemical data (RIVM 2001). For plant protection products and biocides, data from registration application packets are used, as well as other relevant data. For other substances, data are drawn from the public literature. A list of data sources is given, which includes on-line databases (e.g., Current Contents, Biosis, Chemical Abstracts, and Toxline), internal databases, handbooks (Mackay et al. 1992, 1993, 1995, 1997, 1999), libraries, and even confidential data (note that USEPA (1985), expressly excludes the use of confidential or privileged data). Data used to derive MPCs must be from original sources (as opposed to review articles, for example). The scope of the literature search must be described, and should reach back to at least 1970. If four or more acceptable chronic studies are available, just a short overview of acute toxicity is acceptable. However, if data from fewer than four chronic studies exist, then all acute toxicity data are evaluated. Both freshwater and marine data are collected, if statistical comparison indicates that they are not different, these data are combined.

In the Danish methodology (Samsoe-Petersen and Pedersen 1995), data are collected from handbooks, databases, and searches of the open literature. Handbooks include ECETOC (European Centre for Ecotoxicology and Toxicology of Chemicals 1993), GESAMP (Group of Experts on the Scientific Aspects of Marine

Protection 1989), Howard (1990, 1991), KemI (Swedish Chemicals Agency 1989), MITI (Ministry of International Trade and Industry, Japan 1992), Nikunen et al. (1990), Roth (1993), and Verschueren (1983). Databases include AQUIRE (1981-present), BIODEG (1992), and LOGKOW (1994). Biodegradability data are estimated using the BIODEG (1992) probability program, when measured data are not available. Literature searches reach back to 1985 and are conducted using BIOSIS. Details of a BIOSIS search profile are given in Annex 2 of Samsoe-Petersen and Pedersen (1995).

For derivation of criteria in Australia and New Zealand (ANZECC and ARMCANZ 2000), data are collected from international criteria documents, the USEPA AQUIRE database, and papers from the open literature that include acute and chronic toxicity data from field, semi-field, and laboratory data, an internal database, and review papers on ecotoxicology. Physical–chemical data are drawn from electronic databases (such as HSDB – Hazardous Substances Data Bank, available via Toxnet at http://toxnet.nlm.nih.gov/) and from Verscheuren (1983; most recent version 2001 CD-ROM) and Hansch et al. (1995). Spanish guidelines (Lepper 2002) specify that published data from all sources may be used to derive criteria. Principal data sources used are on-line databases (e.g., AQUIRE, POLTOX, and MEDLINE) and published WQOs.

In the UK, data for EQS derivation are taken from published literature, commercial databases, and unpublished sources (such as manufacturer data; Zabel and Cole 1999). The Canadian guidelines (CCME 1999) indicate what kinds of data should be sought, but do not specify data sources. OECD (1995), German (BMU 2001; Irmer et al. 1995), USEPA (1985), EU (ECB 2003; Bro-Rasmussen 1994), France (Lepper 2002), and South African (Roux et al. 1996) guidelines contain no specifics regarding where to find data or what constitutes an adequate literature search.

Without specific requirements for data sources and literature searches, data sets used in criteria derivation could be unnecessarily biased (unnecessary because acceptable data may be overlooked). To ensure inclusion of all relevant data, specific guidance should be part of a derivation methodology.

6.2 Data Quality

To minimize uncertainty in water quality criteria, only data that meet stated quality standards should be used in criteria derivation. Toxicity and physical–chemical data should be from studies conducted according to accepted protocols that are appropriate for the chemical and the organism being tested. All of the reviewed criteria derivation methodologies have specific data quality requirements for physical–chemical data, as well as for ecotoxicity data. In terms of quality, some, such as France, Germany, and Spain, simply state that tests have to be conducted according to accepted, standardized protocols, or according to principles of good laboratory practice (Lepper 2002; BMU 2001; Irmer et al. 1995). Others list very specific data requirements, examples of which are discussed below.

Only a few of the guidelines give specific data quality parameters for some kinds of physical–chemical data. The Dutch methodology (RIVM 2001) requires that solid–water partition coefficients (K_p) be determined in batch experiments as described in Bockting et al. (1993). Tests conducted according to the OECD guidelines are also acceptable. The Netherlands guidance also points out that water solubility should be determined at an appropriate temperature, usually at 25°C, which matches standard laboratory toxicity test temperatures. Because other physical–chemical parameters such as vapor pressure, Henry's constant (K_H), octanol–water partition coefficient (K_{ow}), and solid–water partition coefficient (K_p) are also temperature-dependent, the temperature at which they were measured should also be noted and values should be adjusted, if necessary (Schwarzenbach et al. 1993).

The OECD guidelines (1995) specify that K_{ow} values may be calculated using the ClogP3 algorithm of Hansch and Leo (1979), or may be taken from the THOR/Starlist database. Both the ClogP3 algorithm and the THOR/Starlist database are now accessible through the Bio-Loom program (Biobyte at www.biobtye.com). For highly hydrophobic compounds (log $K_{ow} > 5$), the OECD methodology (1995) requires that the K_{ow} be determined by either the slow stirring or generator column method. The guidelines recommend expert evaluation of K_{ow} values, because there are many compounds for which reliable values cannot be determined. If measured data are not available, OECD (1995) allows that water solubility may be determined by appropriate QSARs that relate K_{ow} to solubility.

The USEPA (1985) has specific criteria for acceptance of bioconcentration factors (BCFs). To be used in determination of final residue values (FRVs), BCFs must be from flow-through tests, must be based on measured concentrations of test substance in both tissue and test solution, and must be from tests that were long enough for the system to reach steady state. For lipophilic materials, the percent lipid in the tissue must be reported. If a BCF was determined at an exposure that caused adverse effects in the test organism, it should not be used. If reported on a dry weight basis, BCF values must be converted to a wet weight basis. Finally, if more than one acceptable BCF is available, the geometric mean of available values is used, provided they are from exposures of the same length.

Any physical–chemical data used in derivation of water quality criteria should be evaluated to ensure that they were determined by appropriate methods. Generally, data from current, standard methods (e.g., ASTM, OECD), applied and performed correctly for the chemical of interest, will be acceptable. Nonstandard methods may also be appropriate, but only if valid reasons are given for deviation from standard methods. In regard to pesticides, which vary widely in characteristics such as hydrophobicity, water solubility, and ionizability, it is particularly important to verify that reported partition coefficients were determined correctly.

For ecotoxicity data, the EU TGD (ECB 2003) defines data quality in terms of reliability and relevance. "Reliability" is the inherent quality of a test, relates to test methodology, and the way that the performance and the results of the test are described. "Relevance" refers to the extent to which a test is appropriate for a particular hazard or risk assessment. "Reliable data" are from studies for which test reports describe the test in detail and indicate that tests were conducted according to generally accepted standards. Relevance is judged by whether a study included appropriate endpoints, was conducted under relevant conditions, and if the substance tested was representative of the substance being assessed. The EU criteria derivation guidance, as described by Bro-Rasmussen (1994), is very general with regard to data quality, and primarily requires that data include details of tests used.

The UK, the Netherlands, Canada, and Australia and New Zealand evaluate ecotoxicity data and assign ratings depending on its reliability and/or relevance. In the UK, primary data are those classified as reliable and relevant, and secondary data are those for which inadequate details are available. The evaluation is based purely on expert assessment of experimental procedures, test species, endpoints, and whether or not a dose–response relationship has been established. Primary data are used in derivation of EQSs; secondary data are used only as supporting information (Zabel and Cole 1999).

The Dutch methodology uses a reliability index (RI) to evaluate ecotoxicity data (RIVM 2001). Reliable data (RI = 1) are from studies conducted and reported in accordance with internationally accepted test guidelines or those set by Mensink et al. (1995). Less reliable data (RI = 2) are those from studies in less accord with accepted guidance or Mensink et al. (1995), and data deemed not reliable (RI = 3) are from studies not at all in accord with accepted guidance or Mensink et al. (1995). Data rated 1 or 2 are used in derivation of ERLs; data rated 3 are included in the final report, but are not used in criteria derivation.

Part of data quality is ensuring that data come from properly conducted, well-documented studies. In the Netherlands (RIVM 2001), data must come from referenced studies that include specific organism identification, information regarding purity of the test substance, details of the test, and clearly stated results. For systematic evaluation, data are subdivided by type (freshwater, marine, acute and chronic) and put into data tables. Table headings should include species (including scientific name), species properties (e.g., age, weight, and lifestage), analysis of test compound (measured or not, Y or N), test type (flow-through, static-renewal, and static), substance purity, test water, pH, water properties (e.g., hardness, salinity), exposure time, test criterion (e.g., LC_{50} – concentration that is lethal to 50% of organisms – or NOEC – no-observed-effect concentration – ecotoxicological endpoint (growth, reproduction, mortality, immobilization, morphological effects, and histopathological effects), LC_{50} values, NOEC values, notes, and reference information.

For data to be usable in criteria derivation in the Netherlands, specific toxicity test acceptability requirements must be met (RIVM 2001). These include that the purity of the test substance must be at least 80%, studies may not use animals collected from polluted sites, concentration of test substance may not exceed 10× the water solubility, no more than 1 ml/L of carrier solvent can be used, and recovery of the substance must be at least 80%. For compounds with short half-lives ($t_{1/2}$), the renewal frequency in a static-renewal test is important. In the Dutch methodology, if the $t_{1/2}$ is shorter than the renewal interval, the data are not used.

The Australia and New Zealand guidelines follow the standard operating procedures for the AQUIRE system (1994) for rating toxicity studies according to how well they are documented (ANZECC and ARMCANZ 2000). In this rating system, weighted scores are applied to 18 characteristics of the test methodology. The two most heavily weighted characteristics are exposure duration and endpoint; both must be recorded for the study to receive a strong rating. Other, lower weighted characteristics, include control type, organism characteristics, chemical analysis method, exposure type, test location, chemical grade, test media, hardness/salinity, alkalinity, dissolved oxygen content, temperature, pH, trend of effect, effect percent, statistical significance, and significance level. Based on scores in these categories, data are rated as C (complete), M (moderate), or I (incomplete). Only data rated C or M are used to derive guideline values. In addition, the Australia/New Zealand guidelines allow for use of data that have already been accepted and used in Dutch and Danish water quality documents. Clear direction on how to handle outlying data is also given in the Australia/New Zealand guidelines, although the curve-fitting technique used in this methodology minimizes the need to remove outliers. The Danish methodology also provides for assessing data quality according to the AQUIRE system (Samsoe-Petersen and Pedersen 1995).

In addition to the Australia/New Zealand data quality guidelines already discussed, the ANZECC and ARMCANZ guidelines (2000) provide specific toxicity test validity criteria. These include that test solutions should cover a geometrically increasing series; a control and solvent control should be included; control mortalities should be less than 10% (or some other level, determined by the specific test method); adverse effects in controls should be less than 20%; water quality parameters should be measured and be within specified limits; a least significant difference for hypothesis tests should be calculated and reported; test organisms should be allowed sufficient time for acclimation to test water; burden of animals in test containers should be appropriate; measured test concentrations should not vary greatly from nominal concentrations; animals should be randomly assigned to test vessels, and test vessels should be randomly placed in test chamber or room; any requirements for things such as timing of hatch, or timing and number of young produced should be met; source and health of test organisms and stock cultures should be traceable; feeding and no–feeding requirements must be met; reference toxicant test results for test organisms should be available.

By the Canadian methodology (CCME 1999), each study is evaluated to ensure that acceptable laboratory practices were used. Studies are classified as primary, secondary, or unacceptable, with only primary and secondary data being used to derive guideline values. Primary data must be from toxicity tests conducted according to currently accepted laboratory practices, but more novel approaches may be acceptable on a case-by-case basis. Also, for primary data, test concentrations must be measured at the beginning and end of the exposure period, and static tests are unacceptable unless test concentrations and environmental conditions were maintained throughout the test. Studies should have endpoints from partial or full life cycle tests and should include determination of effects on embryonic development, hatching, germination, survival, growth, and reproduction. Appropriate controls must be included and measurements of abiotic variables (e.g., temperature, and pH) should be reported. Secondary data may come from tests conducted from a wider range of methodologies, and may include static tests, and tests with endpoints such as pathological, behavioral, or physiological effects. Nominal test concentrations are acceptable for secondary data, and, as for primary data, relevant abiotic variables and control responses should be reported.

Data used in derivation of criteria by the USEPA (1985) must be available in a publication or be in the form of a typed, dated, and signed document (manuscript, memo, letter, etc.). Reports must include enough supporting information to indicate that acceptable test procedures were used and reliable results were obtained. The USEPA also provides very specific data quality guidance (USEPA 1985). Tests are to be rejected if there was no control treatment, if too many control organisms died, were stressed, diseased, or if improper dilution water was used. Tests using formulated mixtures or emulsifiable concentrates are not acceptable, but tests with technical grade materials are acceptable. For highly volatile or degradable materials, or for measuring chronic toxicity, flow-through tests, with frequent measurements of test solution concentrations should be used. Chronic test exposures may continue through the entire life cycle, partial life cycle, or early life stage. Data are rejected if tests were conducted with brine shrimp, a species that does not have reproducing populations in North America, or organisms previously exposed to contaminants.

As in the USEPA methodology, South Africa (Roux et al. 1996) rejects data from tests in which there was no control treatment, there was excess death of control organisms (>10%), there was improper dilution of test water, organisms were previously exposed to contaminants, or where there was insufficient agreement of toxicity data within and between species. Data from tests with formulated mixtures are also rejected.

The OECD guidelines (1995) prefer toxicity data from tests conducted according to standardized methods. The guidelines also specify that it is important to consider water solubility, K_{ow}, and bioaccumulation potential of a substance, when assessing the probable acceptability of acute toxicity data. If the water solubility of the test substance is below the LC_{50}, or if the test duration was too short in the context of the K_{ow} and/or BCF (generally for log $K_{ow} > 5$), then acute tests are not acceptable and only chronic data may be used. Further toxicity test acceptability requirements are not given in OECD (1995), but OECD test guidelines include specific validity standards (e.g., OECD 1992).

6.3 Data Quantity–Ecotoxicity

Small ecotoxicity data sets are a common problem faced by regulators wanting to develop water quality criteria. Large, acute, and chronic data sets, representing numerous taxa, exist for few chemicals. Basic data sets required for pesticide registration, typically comprising only acute data for a few species, are available for many chemicals, but no ecotoxicity data are readily available for some new chemicals. This section explores the kinds of water quality criteria that can be derived from data sets of all sizes.

The quantity of ecotoxicological effects data required for criteria derivation is substantially different for different countries, and depends on what derivation methodology is used, what type of criterion is being developed (i.e., values to be used in standard setting vs advisory values), and what level of uncertainty is acceptable in the criterion. Criteria are derived by extrapolating from available effects data to real-world situations. The two basic methods for doing these extrapolations are (1) application of AFs and, (2) statistical extrapolation of SSDs. There is not much debate about what constitutes appropriate levels of data for the AF method. Factors are applied according to the types and volume of data available, and many methodologies allow for derivation of a numerical guideline value (as opposed to an enforceable criterion). Such guideline values for a contaminant may be based on as little as one datum and may be an estimated toxicity value (e.g., calculated from a QSAR) rather than a measured one. In contrast, for methods that utilize statistical extrapolation, there is little agreement among methodologies concerning how much data are needed to produce criteria with the appropriate level of certainty.

High reliability TVs, as defined in the Australia/New Zealand methodology (ANZECC and ARMCANZ 2000), can be determined either directly from at least three multispecies chronic NOEC values or from statistical extrapolation using at least five single-species chronic NOEC values (from five different species). A moderate reliability TV can be derived from at least five single-species acute toxicity values, and a low reliability TV can be derived from a single acute or chronic toxicity datum.

The Dutch methodology (RIVM 2001) requires at least four chronic NOEC values from species of different taxa for a refined effects assessment; for a preliminary effect assessment, an ERL may be derived from a single LC_{50} or QSAR estimate. Toxicity values, estimated by QSARs, may also be used in statistical extrapolation models.

The OECD guidelines (1995) present several methods for criteria derivation, and each has its own data requirements. For statistical extrapolations using the methods of Aldenberg and Slob (1993) or Wagner and Løkke (1991), at least five chronic NOECs are required. To derive a final chronic value (FCV) using the USEPA methodology (USEPA 1985) requires chronic NOEC values for at least eight animal families, including Salmonidae, a second family in the class Osteichthyes, a family in the phylum Chordata, a planktonic crustacean, a benthic crustacean, an insect, a family in a phylum other than Arthropoda or Chordata, and a family in any order of insects or any phylum not yet represented. Unlike the USEPA method (1985), the OECD does not allow for derivation of a chronic criterion by application of an acute-to-chronic ratio (ACR) to a final acute value (FAV). An environmental concern level (ECL) can be determined by the OECD AF method (1995) from a single LC_{50} value. If no toxicity data are available, QSARs may be used to estimate toxicity for some classes of chemicals. Such estimated values may be used to derive MTCs.

For derivation of a FAV, the USEPA (1985) requires acute toxicity data for eight North American species, each of which must represent different families, as previously described for the OECD methodology. Despite having fewer than the required eight families represented in the data set, the California Department of Fish and Game has derived criteria for carbaryl [1-naphthalenyl methylcarbamate] and methomyl [methyl *N*-[[(methylamino)carbonyl]oxy]ethanimidothioate] using the USEPA SSD method (1985). In so doing, professional judgment was used to determine that species in the missing categories were relatively insensitive, and their addition would not lower the criteria (Siepmann and Jones 1998; Monconi and Beckman 1996). An FCV may be calculated (USEPA 1985) in the same manner as is the FAV, if chronic data are available for at least eight different families. Alternatively, an FCV may be derived by application of an ACR to an FAV, if ACRs are available for aquatic species in at least three families (one of which must be a fish, one an invertebrate, and one an acutely sensitive freshwater species). The USEPA methodology also requires data from at least one toxicity test with an alga or vascular plant, and at least one acceptable BCF. The South African methodology (Roux et al. 1996) has essentially the same data quantity requirements as does the USEPA, with the exception that the data must be from species that are either indigenous to southern Africa, or are of local commercial or recreational importance.

The state of North Carolina follows USEPA FAV derivation procedures (1985) to determine acceptable acute toxicity levels and also provides a means for derivation of an acceptable level of acute or chronic toxicity based on the lowest available LC_{50} value, implying that a single value may be used (North Carolina Department of Environment and Natural Resources 2003). The water quality guidance for the Great Lakes (USEPA 2003a) allows for derivation of Tier II criteria based on applying an AF to the lowest genus mean acute value (GMAV) in the database. Although not explicitly stated, it appears that a Tier II criterion could be based on a single datum, using this method.

The Canadian methodology (CCME 1999) requires studies on at least three fish species resident in North America, including at least one cold- and one warm-water species. A minimum of two of the fish studies must be chronic studies. The Canadian guidelines also require two chronic studies on at least two invertebrate species from different classes, one of which must be a planktonic species resident in North America. A minimum of one study of a freshwater vascular plant or algal species resident in North America is also required, unless a chemical is known to be highly phytotoxic; in that case, at least four acute and/or chronic studies of nontarget plants or algae are required.

For effects assessed according to the EU TGD on risk assessment (ECB 2003), an AF method is used to derive a PNEC. To accomplish this, either one LC_{50}/EC_{50} (lethal or effect concentration to kill/produce effects in half of a tested population, respectively) from each of three trophic levels (fish, crustacean, and alga), or one or more chronic NOECs, are used. Using statistical extrapolation (SSD method), the TGD requires at least ten chronic NOECs from eight taxonomic groups, as follows: two families of fish, a crustacean, an insect, a family in a phylum other than Arthropoda or Chordata, a family in any order of insect or any phylum not already represented, and, an alga and a higher plant.

In France, data from three trophic levels (algae/plants, invertebrates, and fish) are required for derivation of threshold values. If data from only two trophic levels are available, provisional thresholds are derived. If there are no data from particularly sensitive species, or if there are data for fewer than two trophic levels, then no criteria are derived (Lepper 2002).

German methodology requires chronic toxicity data from four trophic levels (bacteria/reducers, green algae/primary producers, small crustaceans/primary consumers, and fish/secondary consumers) to derive criteria. If chronic NOECs are available for at least two trophic levels, acute data may be used to fill trophic level gaps, but must be multiplied by an acute-to-chronic extrapolation factor (0.1); this result is regarded to be a tentative criterion. If chronic data from at least two trophic levels are not available, no criterion can be derived (Lepper 2002; BMU 2001; Irmer et al. 1995).

In Spain, aquatic life criteria are derived from acute or chronic data for at least three species; species must include algae, invertebrates, and fish (Lepper 2002). The UK requires acute or chronic data for algae or macrophytes, arthropods, nonarthropod invertebrates, and fish to derive aquatic life criteria (Zabel and Cole 1999). Neither of these methodologies describes how much data of each kind is required.

Several current derivation methodologies allow water quality guideline values to be derived by applying AFs, even if experimental toxicity data are absent (derivations are based on QSARs). If enforceable criteria, which can be used directly in setting water quality standards, are sought, a large, diverse ecotoxicity database is required. The Canadian guidelines (CCME 1999) require at least six types of data; others do not specify a number, but leave much to professional judgment. In all cases, as the number and diversity of data increase, AFs decrease, thus reducing the uncertainty-driven conservatism in criteria values.

For statistical extrapolations by parametric techniques, data requirements range from $n = 4$ to 10. In discussing the use of statistical extrapolations for very small data sets, Aldenberg and Luttik (2002) noted that sample sizes as small as $n = 2$ can be used; however, the values derived from samples as small as $n = 2$–3 are not of much practical use because of their very high level of uncertainty. Wheeler et al. (2002) analyzed the influence of data quantity and quality, and model choice on SSD outcomes. They found that a minimum of $n = 10$ was required to obtain a reliable estimate of a particular endpoint (e.g., an HC_5; hazardous concentration potentially harmful to 5% of species). Okkerman et al. (1991) conclude that, although seven kinds of data would be ideal, five are adequate for the SSD procedure, described by Van Straalen and Denneman (1989). According to Aldenberg and Slob (1993), the risk that a 50th percentile confidence limit estimate of the HC_5 will result in underprotection decreases considerably as sample size is increased from 2 to 5, but less so as it is increased from 5 to 10 and from 10 to 20.

Jagoe and Newman (1997) proposed using bootstrapping techniques with SSDs to avoid the issue of fitting available data to a particular distribution. Later, Newman et al. (2000) found that the minimum sample sizes required for a bootstrapping method ranged from 15 to 55. In a similar analysis, Newman et al. (2002) discovered that 40–60 samples were required to derive an HC_5 with an acceptable level of precision. Van Der Hoeven (2001) described a nonparametric SSD method that requires a minimum of 19 samples, with as many 59 required to derive a one-sided 95% confidence limit HC_5 estimate. Considering the general dearth of ecotoxicity data, it is understandable that none of the current criteria derivation methodologies utilize a bootstrapping approach for SSD extrapolations. Moreover, Grist et al. (2002) argued that a drawback of using the bootstrap technique is that there is no legitimate way to determine a minimum sample size.

Based on the foregoing discussion, a sample size of 5 is the minimum needed for applying parametric statistical extrapolation procedures. For smaller data sets, only AF derivation methods are appropriate. Minimal data sets, available in the US for derivation of pesticide criteria, are those required for registration under the Federal Insecticide, Fungicide, and Rodenticide Act (FIFRA), and those required by individual states. According to 40 CFR (Code of Federal Regulations) Part 158.490 (USEPA 1993), the minimum data required by FIFRA is an LC_{50} for a fish and an LC_{50} for a freshwater invertebrate. All other kinds of aquatic toxicity data are only conditionally required, but depend on an evaluation of the following parameters: nature of the planned pesticide usage, potential for transport to water, whether acute LC_{50}/EC_{50} values were less than 1 mg/L, whether estimated environmental concentrations are greater than 0.01 times any LC_{50}/EC_{50}, or if data provide evidence of reproductive toxicity, persistence, or bioaccumulative potential. It is possible that many new chemicals will only have the results of two acute toxicity studies available. In California, the Department of Pesticide Regulation (DPR) has tiered data requirements (California DPR 2005a). The minimum data set includes LC_{50}s for one warm-water and one cold-water fish and for a freshwater invertebrate. Further testing is required using the same rationales as those discussed for FIFRA. Again, for new pesticides, it is possible that no more than the minimum data will be available for criteria derivation. To allow regulators to derive criteria for pesticides, based only on the limited data sets submitted for registration, an AF method is needed.

6.4 *Kinds of Data*

Physical–chemical data are not used directly in the derivation of water quality criteria. However, they are valuable for (1) assessment of toxicity test data (e.g., comparing test concentrations to solubility parameters); (2) translation of criteria that are based on total concentration in water to criteria based on dissolved concentration in water or concentrations of suspended matter; (3) assessment of factors that might affect toxicity (such as the effect of pH on the relative concentrations of ionized and unionized forms of chemicals); (4) estimation of physical–chemical parameters for which no measured values are available; (5) prediction of bioaccumulation or secondary poisoning potential, and especially; (6) estimation of toxicity where data are lacking. Herein, BCF and bioaccumulation factors (BAFs) are considered physical–chemical parameters, although it is recognized that they may also be associated with toxicological data.

6.4.1 Physical–Chemical Data

The Netherlands methodology (RIVM 2001) requires collection of specific physical–chemical data. For each substance the following information is required: IUPAC name (International Union of Pure and Applied Chemistry), CAS number (Chemical Abstract Service), EINECS number (European Inventory of Existing Commercial Substances), structural formula (including diagram), empirical formula, molar mass, octanol–water partition coefficient (K_{ow}), water solubility, melting point, vapor pressure, Henry's law constant (K_H), acid dissociation constant(s) (pK_a), solid–water partition coefficients (K_p) and degradation information (i.e., hydrolysis, photolysis, and biodegradation). The methodology includes procedures for calculation of a dimensionless K_H, if experimentally derived constants are not available.

Physical–chemical data and environmental fate information are used in the Dutch methodology (RIVM 2001) in several ways. For example, if a substance has a $t_{1/2}$ of less than 4 hr, then the criterion is derived for stable degradation products, rather than for the parent compound. Moreover, if data are lacking for a particular environmental compartment, partitioning data can be used to estimate concentrations in another compartment. Water solubility data are used to judge the reliability of aquatic toxicity studies, but may also be used together with vapor pressure and molecular weight data to calculate a Henry's Law constant. Suspended matter–water partition coefficients are used to calculate total toxicant concentrations in water, based on the dissolved concentrations. K_{ow}s are used to estimate aquatic toxicity using QSARs, and for estimation of BCF values. K_{ow} values may also be used to determine the potential risk of secondary poisoning, and for estimating organic carbon–water partition coefficients. Finally, partitioning constants are used for harmonization procedures.

The OECD methodology (1995) recommends that the following information be obtained for each compound: chemical structure, molecular weight, melting point, water solubility, K_{ow}, sediment–water partition coefficient (K_{sw}), and pK_a. K_{ow} may be used to estimate water solubility, or to derive QSAR estimates of toxicity. Van Leeuwen et al. (1992) showed that by using QSAR estimates, it is possible to develop a relationship between K_{ow} and the hazardous concentration for nonpolar narcotic chemicals; thus, it is possible to derive MTCs, and their associated confidence limits, directly from K_{ow} values.

In the Australia/New Zealand guidelines (ANZECC and ARMCANZ 2000), K_{ow} and BCF values are used to estimate bioaccumulative potential. The BCF also may be used to calculate water concentrations that will protect fish-eating predators from bioaccumulative chemicals. To derive low reliability target values for narcotic chemicals (when little-to-no toxicity data are available), the Australia/New Zealand guidelines utilize K_{ow} values to derive (QSAR) estimates of toxicity. Beyond K_{ow} and BCF values, the Australia/New Zealand guidelines provide no specific requirements for collection and reporting of physical–chemical data.

For compounds with partition coefficients greater than 1000 L/kg, the German derivation methodology utilizes the solid–water partition coefficient to express quality targets, in terms of toxicant concentration in suspended particulate matter. To protect German fisheries, BCF values are also used to derive water quality targets for fish from pesticide MPR (maximum permissible residue) values.

The USEPA guidelines (1985) require collection of bioaccumulation data only when residues are known to be of toxicological concern. Physical–chemical data, such as volatility, solubility, and degradability, are required for evaluation of toxicity data. BCFs and BAFs are used to derive the FRV.

For development of a full guideline, Canada (CCME 1999) requires collection of environmental fate data. Specifically, information must be available on the mobility of the substance and its final disposition, abiotic and biotic transformations that occur during transport and after deposition, the final chemical form of the substance, and on the persistence of the substance in water, sediment, and biota.

The Danish methodology (Samsoe-Petersen and Pedersen 1995) does not clearly specify what kinds of physical–chemical data must be collected, but criteria derivation documents indicate consideration of a wide-range data. Such data include CAS number, empirical formula, molecular weight, water solubility, K_H, BCF, and K_{ow}, as well as biodegradability data. Bioaccumulation data are used to determine the size of the AF to be applied. Biodegradation data are used to determine whether criteria ought to be derived for the parent chemical or for a stable, toxic metabolite. If little is known about degradation products, then AFs will reflect this uncertainty.

According to EU guidance (Bro-Rasmussen et al. 1994), physical–chemical data requirements are very general, and simply state that "a summary of the main chemical and physicochemical characteristics" must be collected. For criteria derivation, bioaccumulative potential and persistence can affect the size of the applied AF. Also, K_{ow} values may be used to derive QSAR estimates of toxicity, when toxicity data are lacking. For assessment of secondary poisoning potential, the EU risk assessment TGD (ECB 2003) utilizes K_{ow} values, adsorption data, hydrolysis and other degradation data, and molecular weight.

Spanish guidelines (Lepper 2002) require collection of physical–chemical data that may have some bearing on the toxicity of the substance. These include speciation, toxicokinetic properties, and relationships between toxicity and water quality parameters. The UK (Zabel and Cole 1999) and South African (Roux et al. 1996) guidelines do not specify particular uses for physical–chemical data in criteria derivation.

Physical–chemical data are used by various methodologies to improve interpretation of ecotoxicity data and to determine whether water quality criteria are set at levels that could potentially harm nonaquatic species (including humans). Without adequate physical–chemical data, it would not be possible to adequately assess potential effects of chemicals. If explicit details, regarding the collection of physical–chemical data, are available, they are important contributors to criteria derivation methodology.

QSARs describe and model mathematical relationships between a chemical's structure and its toxicity. According to Jaworska et al. (2003) QSARs are simplified mathematic representations of complex chemical–biological interactions. They are most commonly developed by regression analysis, neural nets, or classification methods (Jaworska et al. 2003). QSARs are used by several existing criteria derivation methodologies to fill data gaps. That is, if little or no toxicity data are available for criteria derivation, toxicity can be estimated using QSARs for some compounds and species.

The most commonly used chemical parameter in QSARs is the K_{ow}. QSARs are developed for classes of chemicals, such as inert, less inert, reactive, and specifically acting chemicals (Verhaar et al. 1992). These classes were described by Vaal et al. (1997b) as nonpolar narcotics, polar narcotics, reactive compounds, and specifically acting compounds. For fathead minnows, Russom et al. (1997) further separated the specifically acting compounds into oxidative phosphorylation uncouplers, acetylcholinesterase inhibitors, respiratory inhibitors, electrophiles/proelectrophiles, and central nervous system seizure agents. Using K_{ow} data alone, QSARs with good predictive power can be developed for narcotic chemicals. However, for chemicals with a specific mode of toxic action, additional physical–chemical data are needed, such as reactivity or pK_a, and the predictive models become more complex (Auer et al. 1990). Ramos et al. (1998) suggest that models based on real phospholipid membrane–water partitioning, rather than K_{ow}s, would more accurately predict the toxicity of polar and nonpolar narcotics. The recent "Workshop on Regulatory Use of (Q)SARs for Human Health and Environmental Endpoints," (summarized in Jaworska et al. 2003) produced a series of papers that provide guidance on assessing reliability, uncertainty, and applicability of QSARs (Eriksson et al. 2003). The papers from this workshop also effectively review the use of QSARs in international decision-making frameworks for prediction of ecological effects and environmental fate of chemicals (Cronin et al. 2003).

When insufficient data are available, several water quality criteria derivation methodologies allow for the use of QSARs to estimate aquatic toxicity (discussed below). When assessing hazards of chemicals for which little or no ecotoxicity data are available, the USEPA Office of Pollution Prevention and Toxics (OPPT) uses QSARs, under the Toxic Substances Control Act (TSCA), to estimate toxicity (Nabholz 1991). Toxicity values, calculated from QSARs, are used in statistical extrapolation or AF methods to derive criteria. In contrast, neither the national (USEPA) nor the newer Great Lakes criteria derivation methodologies allow the use of QSARs in criteria derivation (USEPA 1985, 2003a).

Although recognizing that QSARs exist for many modes of toxic action, the Dutch guidelines allow the use of QSARs only for substances that have a nonspecific mode of action (i.e., those acting by narcosis; RIVM 2001). The guidelines provide 19 QSARs for aquatic species that represent nine different taxa. NOECs estimated from QSARs may be used as inputs into extrapolation models for derivation of ERLs. In the UK (Zabel and Cole 1999) QSARs, or other models may be used to predict toxicity in the absence of other data, but such data are not used to derive EQSs (used only for support).

The OECD guidelines (1995) offer two QSAR approaches. First, is borrowed from the USEPA's OPPT, and is based on the classification of chemicals by their structure, without considering mode of toxic action. The specifics of this approach are described by Nabholz (2003). Second, is a method that classifies chemicals, first by mode of action, and then by chemical structure. The second approach is similar to that used in the Dutch methodology (RIVM 2001), except that the OECD provides QSARs for four classes of toxic modes of action (inert/baseline, less inert, reactive, and specifically acting chemicals), as defined by Verhaar et al. (1992). When using the OECD methodology (1995), if no toxicity data are available, QSARs may be used to derive MTCs. For inert chemicals, QSARs may be used to estimate toxicity for fish, *Daphnia* and algae. For chemicals that are not inert, estimates may be made for fish. QSARs are not used to derive OECD MTCs for reactive and specifically acting chemicals, because its application for this purpose has not been adequately evaluated (OECD 1995). If some toxicity data are available, then QSAR estimates of toxicity for inert chemicals are compared to experimental values. If the values agree within a factor of 5, then the QSAR values may be used to extend the database for MTC derivation. MTCs derived solely from QSAR data are used only for priority setting purposes; they are not used to set EQSs.

When toxicity data are unavailable, QSARs may be used to estimate toxicity and fill data gaps for polar and nonpolar narcotic chemicals. However, existing criteria derivation methodologies do not endorse the use of QSARs to estimate the toxicity for chemicals with specific modes of action.

6.4.2 Ecotoxicity Data

Many types of ecotoxicity data exist in the literature. Results of short-term acute and long-term chronic tests are available, as are tests on sensitive life stages. The data from such tests may be used as predictors of chronic toxicity (USEPA 2002b). Some studies are designed to assess lethality, while others address sublethal endpoints, including inhibition of growth or reproduction. Results of other studies are used to evaluate effects of toxicants on biochemical endpoints, such as inhibition of acetylcholinesterase or upregulation of glutathione *S*-transferases. Some tests are performed on only one species, while others simultaneously test multiple species in microcosms or mesocosms. Some tests, performed under laboratory conditions, are tightly controlled, and others are conducted in field or semi-field settings. Each study type generates a value, or series of values, such as LC/EC_x or NOEC. Results may also be reported as a lowest observed effect concentration (LOEC), the lowest concentration of toxicant that causes a response that *is* different from the control, or a maximum allowable toxicant concentration (MATC), which is the geometric mean of the NOEC and LOEC values (USEPA 1987). The following discussion addresses the many different approaches employed to define and use different kinds of data among existing criteria derivation methodologies.

Water quality criteria need to protect aquatic life exposed for short or long periods, and those either transiently or continuously exposed. Long-term exposures are generally considered chronic exposures, while short-term exposures are considered acute. However, a relatively short exposure to an organism with a relatively short life span may be equivalent to a chronic study. What constitutes acute or chronic durations for toxicity tests varies with the test species. Thus, clear guidance on the kinds of (acute and chronic) toxicity test data needed is important, and should be considered in the context of what criteria may be derived from them.

The Netherlands guidelines define acute exposure as lasting a relatively short period; chronic exposures continue through a complete or partial life cycle. Whether an exposure is acute or chronic depends on the physiology and life-cycle characteristics of the species (RIVM 2001). The Dutch guidelines further clarify that acute tests generally last less than 4 d, and results are reported as an LC_{50} or EC_{50}. Chronic tests generally last more than 4 d, and results are reported as an NOEC. However, for single-celled organisms (e.g., algae or bacteria), chronic NOECs may be obtained in less than 4 d. In addition, the guidelines are very specific in stating the following: for algae, bacteria, or protozoa, tests of 3–4 d are defined as chronic; for Crustacea and Insecta, tests of 48 or 96 hr are acute; and, for Pisces, Mollusca, and Amphibia, tests of 96 hr are acute, while early life stage tests and 28 d growth tests are chronic (RIVM 2001). Only chronic NOECs are used for refined effect assessments; acute data are used with application of AFs in preliminary effect assessments.

Chronic toxicity data are preferred by OECD guidelines (1995), with acceptable data presented either as NOECs or MATCs. However, acute data are also used, but with appropriate application of AFs (i.e., ACRs). The guidelines caution that substances with low water solubility, or log $K_{ow} > 5$, a 96 hr acute exposure in water, may not be long enough to see effects, and, therefore, only chronic data may be appropriately used for such substances. Although not explicit, it appears that this methodology considers exposures longer than 96 hr to be chronic. By this methodology, NOECs may be estimated by conversion from LOEC values (e.g., NOEC = LOEC/2), but only if the LOEC corresponds to a concentration causing an effect more than 20%.

The Australia/New Zealand guidelines (ANZECC and ARMCANZ 2000) contain the general description that acute tests are shorter than chronic tests. They proceed to say that in applying the methodology, data from tests longer that 96 hr are considered to be chronic, except for tests with single-celled organisms (for which 96 hr tests are considered chronic). Chronic data are used to derive high reliability target values, while acute data are used to derive moderate reliability target values. NOEC and LC_{50} data are both used in statistical extrapolations, but the resulting hazardous concentration determined with LC_{50} s is multiplied by an ACR.

The USEPA methodology (1985) utilizes acute LC_{50} or EC_{50} data to derive the FAV. The EC_{50} data, in this case, are based on the percentage killed plus the percentage immobilized. EC_{50} data, relating to less severe effects, are not used to calculate the FAV. Acute toxicity data are described as those from 48-h tests with daphnids and other cladocerans, from 96 hr tests with embryos and larvae of various shellfish species, or from 96 hr tests with older life stages of shellfish species. Tests with single-celled organisms, of any duration, are not considered to be acute tests. Expanding on this point, the Great Lakes guidance (USEPA 2003a) states that any test that reports results giving the number of young produced (e.g., protozoan tests) are not considered acute, even if the test duration is less than 96 hr. The USEPA guidance (USEPA 1985) considers the following types of tests to be chronic:

life-cycle tests (those ranging from just over 7 d, for mysids, to 15 mon for salmonids); partial life-cycle tests (those with all major life stages exposed, but less than 15 mon in duration; specifically for fish that require more than 1 yr to reach sexual maturity), or those that test early life stages (those that range from 28 d to 60 d; particularly applies to fish). Chronic data (reported as an MATC value or determined by regression analysis) are used to derive an FCV. The FCV may also be calculated by applying an ACR to the FAV. The South African methodology (Roux et al. 1996) generally follows that of the USEPA (1985), in terms of using LC_{50}/EC_{50} and MATC data, but does not contain explicit descriptions of acute versus chronic data.

The German guidelines (Irmer et al. 1995) use NOECs derived from studies of long-term toxicant exposure. If no chronic data are available, acute data may be multiplied by a factor, and used instead. No guidance is given on how to classify tests as either acute or chronic.

For derivation of full guidelines, in Canada (CCME 1999), the results of at least two of three fish toxicity studies must be from full or partial life-cycle (chronic) studies, and both invertebrate studies must be from full or partial life-cycle exposures. Rather than using NOEC values, as most methodologies do, the Canadian methodology uses the lowest observable effect level (LOEL; equivalent to LOEC), to derive guidelines. ACRs may be used to convert acute data to chronic values. A study on plants, of unspecified duration, is required for most substances. However, for highly phytotoxic substances, four acute and/or chronic studies are required (duration of acute vs chronic is not defined for plants).

The UK guidelines (Zabel and Cole 1999) use both acute and chronic data for derivation of annual average (AA) concentrations, but use only acute data for derivation of maximum allowable concentrations (MAC). Chronic values may constitute chronic or subchronic NOECs, MATCs, or chronic EC_{50}s. No guidance is given on how to distinguish between acute and chronic data for nonplants, although the guidelines specify that algal growth tests lasting 48–72 hr represent chronic exposures, and should not be used to derive an MAC. However, they state that tests measuring algicidal effects in a 48–72 hr exposure would be appropriate for derivation of an MAC. If, however, algae are the most sensitive of species tested for a substance, then a growth inhibition EC_{50} may be used to derive an MAC.

Both acute LC_{50} and chronic NOEC data may be used according to the EU methodology (Bro-Rasmussen 1994), but definitions for what constitutes acute or chronic are not given. The EU risk assessment TGD (ECB 2003) avoids the use of the terms "acute" and "chronic" and, instead, refers to short- and long-term tests. Short-term results are in the form of LC_{50}s/EC_{50}s, and long-term results are in the form of NOECs, which may be estimated from LOECs, EC_{10}s, or MATCs. The only guidance given, regarding what duration constitutes a short- versus long-term exposure is for algae studies; such studies are considered to be short-term if they are less than 72 hr, and long-term if they are 72 hr or longer.

The Danish, French, and German guidelines (Samsoe-Petersen and Pedersen 1995; Lepper 2002) all utilize both LC_{50} and NOEC data to derive criteria, but none of them specifically define acute versus chronic tests.

To ensure consistency in how toxicity data are used to derive criteria, the terms "acute" and "chronic" must be defined in the methodology. Once defined, the choice to use either acute or chronic data depends on what kind of criterion is being calculated and what kinds of data to support it are available. Acute criteria should be derived from acute data, and chronic criteria should be derived from chronic data; however, when chronic toxicity data are lacking, acute data may be used to derive chronic criteria.

As discussed in other parts of this chapter, current criteria derivation methodologies use toxicity data that have been summarized in the form of an NOEC, LC_{50}, EC_{50}, or some other effect level (i.e., EC_5, EC_{10}, EC_x, etc.). Which, then, among these values is best to use for derivation of protective criteria? We address this question in the following section. It comprises a discussion of toxicity data analysis methods, which focus on the problems and challenges associated with using either NOEC or EC_x values for deriving protective criteria.

Ecotoxicity test data are usually analyzed by one of two methods: hypothesis tests or regression analysis. Hypothesis tests are typically used to analyze results of life cycle, partial life cycle, and early life-stage tests. In this approach, results from treatment groups are compared with those from control groups to determine significant differences in responses (Stephan and Rogers 1985). An NOEC or NOEL (no-observed-effect level), and an LOEC or LOEL may be derived from this type of analysis. Some methodologies use the geometric mean of the NOEC and the LOEC to calculate an MATC. The other widely used method for analysis of ecotoxicity data is regression analysis, which is most commonly applied to acute toxicity tests, but can as easily be applied to chronic tests. In regression analysis, an equation is derived that describes the relationship between concentrations and effects (Stephan and Rogers 1985). Thus, it is possible to make point estimates of toxicant concentrations that will cause a given level of effect (EC_x), or to predict effects for a given level of toxicant.

Many problems with the hypothesis test approach are described in the literature. They are summed up succinctly by Stephan and Rogers (1985) who point out seven computational and five conceptual problems with hypothesis testing, and then discuss why regression analysis is a better alternative. The computational points are briefly described here; for the conceptual points and further details, the reader is referred to Stephan and Rogers (1985):

1. Hypothesis testing can only provide quantitative information about toxicant concentrations actually tested. The estimated effect values (i.e., NOEC and LOEC) must be one of the tested concentrations, with the true NOEC lying somewhere between the NOEC and the LOEC. For regulatory purposes, such as deriving water quality criteria, a single number is needed; therefore, regulators choose to use one or the other of the NOEC or LOEC, or they use an arithmetic or geometric mean of the NOEC and the LOEC. As the authors point out, hypothesis tests provide no basis for such interpolations. In contrast, regression analysis determines a relationship between concentration and effect, and so provides a means to interpolate for estimation of effects at untested concentrations.

2. Hypothesis tests are sensitive to how carefully a test was conducted (i.e., a well-conducted test typically produces low variability within treatments), and how many replicates were used. In other words, the minimum detectable significant difference between treatments decreases with increased replication and with decreasing variability between replicates. In regression analysis, the point estimate is not affected by the number of replicates or the reproducibility among replicates; only the size of the confidence limits is affected.
3. In hypothesis testing, the selection of α (type I error rate), which is usually arbitrarily chosen at 0.05, can completely change the resulting NOEC value. With regression analysis, the confidence limits will change according to α, but the point estimate will not change.
4. The effect value obtained from a hypothesis test is completely dependent on what toxicant concentrations were actually tested. Regression analysis allows for estimation of a concentration that falls between those actually tested. Consequently, regression analysis provides a way to predict an effect level for any given concentration, which cannot be done with the results of hypothesis tests.
5. Changes in statistical procedure (such as use of data transformations) can have large effects on results of hypothesis tests because of the discontinuous nature of the data. For example, if the results of a hypothesis test are changed by a data transformation, the change in the resulting effect level will probably be at least a factor of 2, which is the reciprocal of the typical dilution factor used in toxicity tests. However, in a regression analysis, the concentration–response curve is assumed to be a smooth continuous function and results are affected very little by small changes in statistical procedures.
6. Hypothesis testing does not properly interpret data inversions. That is, if a particular toxicant concentration caused a significant effect, but a higher concentration in the same test did not, then interpretation of hypothesis test results is difficult. The same result, when analyzed by the regression approach, would just widen the confidence limits of the point estimate.
7. Hypothesis tests require averaging of experimental units across replicates. For example, if measured concentrations for a particular treatment vary, then the concentrations must be averaged before the hypothesis test can be conducted. With regression analysis, each experimental unit can be treated independently. If concentrations vary within intended replicates, the results can be used without averaging.

The most important conceptual point made by Stephan and Rogers (1985) is that hypothesis tests give results that are statistically significant, but have nothing to do with the biological significance of effects. Hypothesis tests are typically performed with the Type I error rate (α) defined, but without proper definition of an acceptable Type II error rate (β), and without specifying an acceptable minimum significant difference. Thus, there is no linkage of statistics to biology. Bruce and Versteeg (1992) observe another shortcoming of hypothesis testing. Namely, when results are reported only as an NOEC value, information on the concentration–response curve and variability in the data is lost.

Hoekstra and Van Ewijk (1993) gave examples of how NOEL values (i.e, no *observable-in-this-particular-test* effect level values) are often misinterpreted as no-effect levels. They cite a study by Murray et al. (1979), in which thymus gland weight was potentially reduced by as much as 25% at the NOEL, with the uncertainty resulting from variability in the weight of the exposed thymus glands. A study by Speijers et al. (1986) resulted in a NOEL that could potentially cause a 73% reduction in response, compared to control values. Mount et al. (2003), similarly noted that tests with low variability may produce an LOEC representing responses 2–3% different from the control, whereas, a test with high variability may produce an LOEC representing responses 40^{+} % different from the control. Stephan and Rogers (1985) found that adverse effects, ranging from 10% to 50% different from controls, have been reported as "no statistically significant effect concentrations." Suter et al. (1987) found effect levels at the MATC, in fish tests, ranging from 12%, for hatching, to 42% for fecundity. In a more recent short communication, Crane and Newman (2000) summarized findings of studies showing that the level of effect corresponding to reported MATCs for fish averaged 28%, with a range of 0.1–84%, and that power analysis of hypothesis tests for standard *Daphnia magna* and *Ceriodaphnia dubia* tests revealed that these tests are able to detect effects ranging from 25% to 100%. Clearly, despite its name, the NOEC is not a no-effect level, and for derivation of protective water quality criteria, it would be unacceptable to use NOEC data corresponding to such potentially high effects.

Given the apparent agreement among toxicologists that regression analysis provides better effect level estimates than hypothesis tests (Stephan and Rogers 1985; Bruce and Versteeg 1992; Grothe et al. 1996; Moore and Caux 1997), we are faced with the problem of having a large, otherwise usable, historical chronic toxicity data set in which results are reported as NOECs derived from hypothesis tests. In some cases (i.e., if enough raw data are included in the study report), data could be reanalyzed to determine point estimates. However, the problem of deciding what effect level best represents a no-effect level, remains. The USEPA (1991) suggests that an NOEC (for all tests and species) is approximately equivalent to an IC_{25} (inhibition concentration; concentration causing 25% inhibition compared to the control), while Bruce and Versteeg (1992) chose an EC_{20} as a level of population effect that probably would not lead to adverse effects at the community level. Bruce and Versteeg (1992) also state that the decision as to what is a safe level should be based on biological criteria established with consideration for the species, the measured endpoint, test design, compound degradability, and the slope of the concentration–response curve. Results of a 1994 workshop in the Netherlands indicated a preference among participants (including regulators, industry, contract laboratories, statisticians, and risk assessors) for use of an EC_5 or EC_{10} to represent a no-effect level (Van Der Hoeven et al. 1997). This was determined via a questionnaire, with responses ranging from EC_1 to EC_{25}. Reasons given for choosing the EC_5 and the EC_{10} were admittedly completely nonscientific: the effect level should be small because an (almost) no-effect level is intended; the effect level should not be too small because of problems with accuracy and model dependence; and the effect level should be a round number.

Participants felt that the effect level should depend on ecological consequences, but that would require species-dependent values when, politically, a single effect value for all species is preferable.

Other, novel ways of analyzing toxicity data have been proposed. These include the use of parametric threshold models to derive a parametric no-effect concentration (parNEC; Van Der Hoeven et al. 1997; Bedaux and Kooijman 1993; Cox 1987), models based on dynamic energy budget (DEB) theory (Kooijman 1993; Kooijman et al. 1996; Kooijman and Bedaux 1996a, b; Péry et al. 2002), the use of life table evaluation techniques (Daniels and Allan 1981; Gentile et al. 1982), case-based reasoning models (Van Den Brink et al. 2002), and the use of a double bootstrap procedure to estimate demographic toxicity (e.g., toxicant effect on population growth rate; Grist et al. 2003). These models are not well developed, and the results they produce have not been thoroughly compared to existing data analysis methods.

A sound approach, then, may be the one proposed by participants in the 1994 workshop in the Netherlands (Van Der Hoeven et al. 1997). There was overwhelming support for replacing the NOEC with a more appropriate measure. However, they recognized the need for a transition period and concluded that NOEC data may be used as a summary statistic in ecotoxicity testing, if the following are reported: (a) the minimum significant difference; (b) the actual observed difference from control; (c) the statistical test used; and (d) the test concentrations. Of the alternative NOEC replacements considered at the workshop, there was no preference for either the EC_x or parNEC approach, because both have merit, and further research is needed before a choice can be made. However, according to workshop participants, if the EC_x approach is used, then the x value should be 5% or 10%.

Statistical regression methods are commonly used and widely accepted for analysis of acute toxicity data. For analysis of chronic data, hypothesis tests have been more widely used, but have fallen out of favor, primarily because of their dependence on experimental design and unrestrained Type II error rates. Regression methods are currently preferred for analysis of chronic data. The problem with regression methods is that they yield EC_5, EC_{10}, or other EC_x values, and science has not yet decided which of those values best represents a true no-effect level. Policy decisions are needed to decide what kind of chronic data are acceptable for use in criteria derivations.

The USEPA methodology (1985) points out that it would be ideal if aquatic no-effect concentrations could be determined by adding various concentrations of a chemical of concern to several clean water bodies, and then determine the highest concentration that causes no effect. Because such an approach is not an option, we must instead rely on smaller-scale toxicity studies, ranging from single- and multispecies laboratory tests to multispecies field or semi-field (microcosm or mesocosm) tests. As models of environmental exposure, the order of preference for obtaining needed results is field tests, followed by mesocosm/microcosm tests, multispecies laboratory tests, and single-species laboratory tests. However, the most abundant, reliable, and easily interpretable toxicity data are from single-species laboratory tests. All of the other types of studies are criticized for lack of standardization, lack of replication, and difficulty of interpretation.

Multispecies data are problematic for use in criteria derivation as a result of its paucity and variability. In contrast, there is much debate in the literature about whether or not single-species toxicity tests are good predictors of ecosystem effects. Schulz and Liess (2001) saw significant differences in fenvalerate [cyano(3-phenoxyphenyl)methyl 4-chloro-α-(1-methylethyl)benzeneacetate] effects on caddisflies from both inter- and intra-specific interactions. However, as previously discussed, single-species toxicity tests can be successfully used in various extrapolation procedures to determine concentrations that are protective of ecosystems (Maltby et al. 2005; Hose and Van Den Brink 2004; Okkerman et al. 1993; Versteeg et al. 1999; Emans et al. 1993; USEPA 1991).

Crane (1997) reviewed use of multispecies, model ecosystem, tests for predicting effects of chemicals in the environment. He concluded that more information is needed on repeatability, reproducibility and predictive ability before such tests can be used confidently for predicting environmental effects. Kraufvelin (1999) studied Baltic Sea hard bottom littoral mesocosms and concluded that repeatability, reproducibility and ecological realism of these mesocosms were poor enough to preclude the use of such data in predictive risk assessment, or for extrapolation to natural ecosystems. Sanderson (2002) reviewed the replicability of micro- and meso-cosms and found that coefficients of variation (CV) averaged 45%, with large, outdoor mesocosms averaging 51%. Also, 88% of biotic variables measured lacked statistically significant results, even with 3–4 replicates that should have yielded results better than reported.

Another problem with field or semi-field studies is that they often have few replicates because of unmanageable logistics. One of the reviewed methodologies (OECD 1995) cites a SETAC-Europe document (1992) that asserts that unreplicated experiments may be acceptable for responses that occur in a short period of time. However, Hanson et al. (2003) found that, to detect a ≥25% change from control values in microcosm exposures of *Myriophyllum* spp. to haloacetic acids, would require between 2 and 21 replicates, depending on what endpoint is measured.

The question of how well single-species toxicity tests predict field effects has been addressed by many researchers. As discussed previously, water quality criteria, derived from single-species tests are protective of ecosystems in many cases. Borthwick et al. (1985) showed that laboratory-derived NOECs were predictive of field effects of fenthion on pink shrimp. Similarly, Crane et al. (1999) found that the response of freshwater amphipods to pirimiphos methyl was the same, whether exposed in 250-mL laboratory beakers or 50,000-L pond mesocosms. However, a caveat to that study is that the amphipods were caged in the mesocosm study and, thereby, did not experience the full effects of the mesocosm environment. While validating field predictions that were based on laboratory-derived NOECs, Persoone and Janssen (1994) concluded that, in general, NOECs derived from single-species laboratory studies relate well to single- and multispecies NOECs derived from field studies.

Field or semi-field data are used in the Dutch methodology for comparison with ERLs derived from single-species data (RIVM 2001), but are not used as input for ERL derivation. Nonetheless, to be usable, the data must meet specific requirements. Studies must show a distinct concentration-effect relationship, derive a reliable multispecies NOEC, include several taxonomic groups, include at least two test concentrations and a control, and include two replicates per concentration. In addition, the concentration of the compound should be measured periodically throughout the study and physical–chemical parameters should be monitored [pH, temperature, hardness, TSS (total suspended solids), etc.]. Test endpoints should include biomass and population density, as well as species diversity and species richness.

Although field and semi-field data are not used in criteria derivation, the OECD guidelines (1995) offer criteria for assessing the acceptability of ecosystem studies; such studies are useful for assessing effects of chemicals under field conditions. To be acceptable, test results must include an NOEC for key components of the ecosystem and must show a concentration–response relationship. To avoid overprediction of toxicity, the test should include an ecosystem recovery component. Test systems should include a range of taxonomic groups, preferably including fish, and must have properly simulated ecosystem properties such as nutrient cycling and trophic structure. Physical and chemical parameters (pH, dissolved oxygen, hardness, and temperature) must be monitored throughout the test. Biological response measurements should include individual level parameters (survival, growth, reproduction, and bioaccumulation) as well as population (age/size structure, production, and recover rates) and community level (species composition and relative abundance) measurements. Tests must be conducted at time and space scales that are appropriate to the physical–chemical characteristics of the toxicant and life history of organisms. Ecosystem studies must include a control and —two to three test concentrations and should be duplicated.

Because of the challenges in conducting and interpreting multispecies toxicity tests, and the relative cost-effectiveness, reproducibility and reliability of single-species tests, most methodologies do not utilize multispecies data for criteria derivation. However, Australia/New Zealand, Germany, the UK, and the EU (in risk assessment TGD) do have provisions for using field or microcosm data to derive criteria, providing it meets acceptability criteria (ANZECC and ARMCANZ 2000; Irmer et al. 1995; Zabel and Cole 1999; ECB 2003). In practice, very few criteria are derived from field data. Methodologies that do not use field or semi-field data directly do use them as a comparison to criteria derived from single-species data (RIVM 2001; OECD 1995). In some cases, a final criterion may be adjusted if strong multispecies evidence indicates that the single-species criterion is over- or under-protective (USEPA 1985, 2003a; Zabel and Cole 1999; RIVM 2001).

Survival, growth, and reproduction are traditional measurement endpoints in ecotoxicity tests. Because these effects can be readily linked to population-level effects, they are favored for use in deriving water quality criteria that are to be protective of ecosystems. Nontraditional endpoints, such as endocrine disruption, enzyme induction, enzyme inhibition, behavioral effects, histological effects, stress protein induction, changes in RNA or DNA levels, mutagenicity, and carcinogenicity, are often more sensitive than traditional endpoints, but have had very few links established between effects seen at the individual versus population, community, or ecosystem levels. For this reason, they are rarely used for derivation of water quality criteria.

Using the USEPA methodology (1985), nontraditional endpoints fall into the category of "other data," and are rarely used in criteria derivation. The recent "Ambient

Aquatic Life Water Quality Criteria for Tributyltin (TBT) –Final" (USEPA 2003b) utilizes data from studies of imposex in the dogwhelk, *Nucella lapillus*, to set the final chronic criterion.

In calculating aquatic ERLs, the Netherlands includes only data from endpoints that affect species at the population level, such as survival, growth, and reproduction (RIVM 2001). However, a broad range of effects is included in the category of "reproductive effects," such as histopathological effects on reproductive organs, spermatogenesis, fertility, pregnancy rate, number of eggs produced, egg fertility, and hatchability (Slooff 1992). Endpoints used for calculation of criteria, based on secondary poisoning, include fertility, pregnancy rate, number of live fetuses, pup mortality, eggshell thinning, egg production, egg fertility, hatchability, and chick survival (Romijn et al. 1993). Other endpoints, such as immobility or endocrine disruption, may be used only as evidence in support of derived ERLs, if the endpoints are relevant to the species or are specific to a known toxicant's mode of action. Carcinogenicity and mutagenicity endpoints are not used because studies of these endpoints are difficult to evaluate. Also, population level consequences from carcinogenicity and mutagenicity are unknown (RIVM 2001).

The OECD methodology (1995) prefers use of traditional endpoints, such as survival, growth, and reproduction; however, biochemical endpoints may be considered as well, although clear guidance is not given for their use. German policy makers have recognized the potential for adverse effects posed by endocrine disruptors, but have not yet incorporated these concerns into water quality targets, because data are unavailable on the presence of endocrine disruptors in water bodies, and on concentration-dependent effects (BMU 2001). Similarly, the Danish methodology (Samsoe-Petersen and Pedersen 1995) excludes data for endpoints such as enzyme activity, or hemoglobin or hormone concentrations because such effects do not translate easily into population level effects. The Canadian methodology (CCME 1999) accepts test results with pathological, behavioral, and physiological effects endpoints; however, such studies are regarded as secondary data (used for derivation of interim guideline values). In South Africa, only lethality (an irreversible effect) is accepted as an acute endpoint (Roux et al. 1996); any adverse effect is accepted as an endpoint for chronic data because such data are consistent with the precautionary approach.

Segner (2005) conducted a review that links exposure and effect for endocrine disruptors. He presented three cases, in which population-level effects in wildlife could be linked to environmental substances with endocrine activity: reductions in dogwhelk (*Nucella lapillus*) populations resulting from imposex caused by exposure to TBT; reduction in predatory bird populations from eggshell thinning caused by exposure to DDE (1-chloro-4-[2,2-dichloro-1-(4-chlorophenyl)ethenyl]benzene); and decline in Atlantic salmon populations from effects of 4-nonylphenol on the ability of smolts to osmoregulate. However, only in the case of TBT is there a strong case for endocrine disruptors actually causing the observed toxic effects. Matthiessen (2000) points out that, despite this result, it is uncertain that endocrine effects in individuals produce population effects for other chemicals—the data just do not exist to draw that conclusion. Triebskorn et al. (2003) undertook a project that incorporated laboratory, semifield and field studies to determine relationships among effects linked to biomarkers, behavior, reproduction, and biomonitoring. The authors found they could extrapolate from biochemical effects to population level effects, but they could not statistically link population and community level responses to chemical, limnological, and geomorphological field data. The causal relationship between environmental conditions and effects could not be established. In addition, several of the enzyme induction responses were not useful indicators of exposure or effects (Triebskorn et al. 2003).

In a thorough review of fish bioaccumulation and biomarkers, Van Der Oost et al. (2003) reported that predicting the extent to which biochemical alterations in a population may influence the health of the population or ecosystem, is difficult. Moreover, they discuss several cases in which fish diseases have been linked to pollutant exposure, and exposure was linked to biomarker response, but they concluded the discussion with further comments about the difficulty of correlating biomarker responses to higher-level responses.

In a study of the effects of atrazine and its degradation products on routine swimming, antipredator responses, resting respiration, and growth in red drum larvae, Del Carmen Alvarez and Fuiman (2005) saw significant effects on swimming behaviors and growth. They also found higher rates of predator–prey interactions. However, the only quantitative prediction they made about population effects resulted from reduced growth rates. The authors postulated that increased metabolic rates, resulting from higher swimming rates, may lead to starvation, but no quantitative link was established.

Others have struggled to understand the significance (to populations, communities or ecosystems) of biomarkers observed in individuals. Olsen et al. (2001) found natural variability as high as twofold in acetylcholinesterase and glutathione *S*-transferase levels in *Chironomus riparius* Meigen larvae exposed at 13 uncontaminated sites. Such variability, in the absence of toxicants, suggests that it is difficult to discern toxicant effects by monitoring activity levels of these enzymes. In a later study, Crane et al. (2002) determined that acetylcholinesterase inhibition in *C. riparius* is a good predictor of demographically important effects (e.g., reproduction), caused by exposure to the insecticide pirimiphos methyl. In the same study, Crane et al. (2002) found that pirimiphos methyl had no effect on glutathione *S*-transferase activity. Callaghan et al. (2002) similarly found that acetylcholinesterase activity in *C. riparius* was a robust and specific biomarker for exposure to organophosphate pesticides, and was unaffected by temperature variation. In contrast, glutathione *S*-transferase activity was neither robust nor specific, with induction occurring at low temperature and in response to pesticide exposure. Enzyme induction was not linked to demographic effects in the study by Callaghan et al. (2002). In a field study of the effects of bleached kraft mill effluents on fish, Kleopper-Sams and Owens (1993) found that induction of P450 enzymes was a good biomarker for exposure, but provided no predictive power for individual health or population level effects.

De Coen and Janssen (2003) have proposed a model for predicting population-level effects based on biomarker responses in *Daphnia magna*. By measuring digestive and metabolic enzyme activities, cellular energy allocation, DNA damage, and antioxidative stress activity, they used a multivariate partial least squares model to predict time to death, mean brood size, mean total young per female, intrinsic rate of natural increase, net reproductive rate, and growth. They found that energy-based biomarker measurements combined with measurements of DNA integrity produced good predictions of population-level effects. Maboeta et al. (2003) found a link between a biomarker and population effects in earthworms. Teh et al. (2005) found that Sacramento splittail suffered reduced survival and growth, as well as cellular stress, after a 3-mon recovery period, following a 96 hr exposure to runoff from orchards treated with diazinon [*O,O*-diethyl *O*-[6-methyl-2-(1-methylethyl)-4-pyrimidinyl] phosphorothioate] and esfenvalerate [(*S*)-cyano(3-phenoxyphenyl)methyl (αS)-4-chloro-α-(1-methylethyl)benzeneacetate]. Although no significant mortality occurred during the 96 hr exposure period, histopathological abnormalities were observed after a 1-week recovery period in clean water. Although it appears that the histopathology may have predicted the population level effects in this case, no mechanistic link was made, and it is possible that the reduced growth resulted from other factors.

Studies showing a predictive relationship between biochemical, behavioral, or other nontraditional endpoints and population, community, or ecosystem level effects are rare. Much more research is needed before nontraditional toxicity test endpoints will be generally useful as predictors of ecosystem no-effect levels.

One criticism of using single-species toxicity data for derivation of water quality criteria is that such tests are performed on a very limited number of species. For the majority of species no toxicity data exists, which can be of particular concern where threatened or endangered species are at risk of chemical exposure. This section presents tools that have potential for using toxicity results from tested species to predict potential for toxicity to untested species.

The concept of quantitative species sensitivity relationships (QSSRs) was developed by Vaal et al. (1997a). They looked for patterns in sensitivity variation among 26 aquatic species for 21 toxicants. Although species could be qualitatively grouped according to sensitivities (e.g., vertebrates were different from invertebrates), no quantitative predictive model could be derived. The authors noted that to further develop QSSRs, their findings need to be interpreted in terms of toxicokinetics, modes of action, and relevant species characteristics. In another study, Vaal et al. (1997b) found that acute lethality of nonpolar and polar narcotics is highly predictable for a broad range of aquatic species. Reactive and specifically acting chemicals tend to be much more toxic, with very high variation in sensitivity between species, and their toxicity is not predictable with current information and models.

The USEPA has developed interspecies correlation estimation software (ICE v 1.0), which can be used to estimate acute toxicity of a wide array of compounds to aquatic species, genera, and families that lack such data (USEPA 2003d). Toxicity estimates made by interspecies correlations work well within taxonomic families, but less well as taxonomic distance increases. The ICE models generate estimated toxicity values, with confidence limits, to quantify uncertainty.

Vaal et al. (1997a, b) does not believe that QSSRs are sufficiently well developed to be generally useful for estimating toxicity to untested species. The EPA ICE model offers a promising technique for generating toxicity estimates for untested species, including threatened or endangered species. Estimates from ICE could be used to supplement data sets so that important untested species may be included in criteria derivations. Estimates may also be used to evaluate whether criteria derived with tested species would protect untested species of particular concern.

6.5 Data Reduction

Data that are to be used in criteria calculation procedures often require preliminary treatment. For example, if there are multiple data for a particular combination of species, substances, and endpoints, some method is needed to reduce those data into a single point for each species/substance/endpoint combination. Most methodologies utilize the geometric mean to represent the best estimate (central tendency) of a toxicity or hazard value, but whether to use the geometric mean or the arithmetic mean for environmental chemical data is somewhat controversial. Parkhurst (1998) argues that, for environmental chemical concentrations, the arithmetic mean is superior to the geometric mean because it is unbiased, easier to calculate, scientifically more meaningful, and more protective of public health (as a result of the low bias of the geometric mean). He acknowledges a few cases in which a geometric mean is preferable. One such case, pertinent to criteria derivation, is that of averaging ratios, such as BCFs. Even for log-normally distributed data, Parkhurst states that the arithmetic mean is preferable, because it is unbiased and makes more scientific sense. He gives the example of two data sets, A = (10, 90) and B = (40, 50). The arithmetic mean of A is larger, but the mean of the logarithms of B is larger. In such a case, according to Parkhurst, a statistical comparison based on the log-transformed data may be irrelevant or misleading.

The USEPA (1985) argues that for log-normally distributed data, the geometric mean is preferred over the arithmetic mean. Parkhurst's argument, regarding the low bias of geometric means not being protective, does not apply to toxicity data (as opposed to environmental concentration data), because lower values are more protective. With regard to Parkhurst's example of sets A and B, the possibility of being misled by the geometric mean is not different than being misled by the arithmetic mean, because the degree of variability between the raw and the log-transformed data are not significantly different.

The USEPA (1985) requires that species mean acute values (SMAVs) are to be calculated as the geometric mean of observed species values; similarly, GMAVs are calculated as the geometric mean of all SMAVs for a given genus. If data indicate that a particular life stage is at least a factor of 2 more resistant than another life stage for the same species, then the data for the more resistant life stage is not used to calculate the SMAV, because the goal is to protect all life stages. Similarly, if acute toxicity values for a species or genus differ by more than a factor of 10, then some or all of the values should be excluded (guidance on how to choose what to keep or exclude is not given). The SMAV may be calculated from the result of one or more flow-through tests, in which toxicant concentrations were measured, but if no such data are available, then data from static or static-renewal tests with nominal toxicant concentrations are used. The same procedure applies to chronic data. Chronic values are calculated from the geometric means of NOEC and LOEC values (i.e., MATC values), or from a value derived by regression analysis (with no indication of whether to use an EC_5, EC_{10}, EC_{25}, or other EC_x value). The South African guidelines follow the USEPA procedure for data reduction (Roux et al. 1996), but specify that only chronic MATCs are to be used for criteria derivation.

The Dutch methodology offers very clear instructions regarding preliminary data processing (RIVM 2001). For a given substance, if several study results are available for one species for the same endpoint, then the geometric mean of these values is used. If study data are available for one species, but have multiple endpoints, then the values for the most sensitive endpoint are used. If data are available for multiple life stages of one species, then data from the most sensitive life stage are used. All acceptable chronic toxicity data are converted into NOEC values, as follows (RIVM 2001):

- The highest reported concentration not statistically different from the control ($p < 0.05$) is the NOEC.
- The highest concentration showing 10% effect, or less is considered the NOEC if statistical evaluation is not possible.
- A reported LOEC is converted to an NOEC by use of factors (factors may be adjusted if justified by data):
  - NOEC = LOEC/2 for cases where: 10% effect <LOEC <20% effect.
  - NOEC = EC_{10} for cases where: LOEC ≥20% effect and dose–response relationship is available.
  - NOEC = LOEC/3 for cases where: 20% <LOEC <50% effect.
  - NOEC = LOEC/10 for cases where: 50≤ LOEC ≤80% effect.
- NOEC is reported as ≥ [highest observed no-effect concentration] if none of the treatment groups was significantly different from the control; these values are not used in statistical extrapolation methods.
- "Toxic Threshold" values, as defined by Bringmann and Kühn (1977) are regarded as NOECs.
- For a MATC expressed as a range of values, the NOEC is the lower value; for MATC expressed as a single value, the NOEC = MATC/2.
- NOEC values expressed as total concentrations in water are converted to dissolved concentrations, if the K_p and concentration of particulate matter are known.

Further data processing is required by the Dutch methodology (RIVM 2001) if toxicity data for a particular toxicant appear to be bimodally distributed; in this case, statistical analysis must be performed to determine if the apparent differences are significant. This requirement would apply to differences between, for example, freshwater and saltwater species. If the differences are not significant, then the data are combined for criteria derivation. If the differences are significant, then separate criteria must be developed.

The EU risk assessment TGD (ECB 2003) also provides instructions for how to derive LC_{50}/EC_{50} or NOEC values from studies, in which those values are not reported:

- If raw data are available, the values can be directly calculated. The LC_{50}/EC_{50} should be calculated by probit analysis, or another regression technique; the NOEC may be calculated using either the hypothesis test or regression approach (the TGD does not claim a preference for one over the other, as a result of the continuing controversy, previously discussed).
- Results presented as $LC_{10–49}/EC_{10–49}$ can be used as LC_{50}s, but if results are presented as $LC >_{50}/EC >_{50}$ they cannot be used.
- An LOEC representing an effect more than 10% and less than 20% may be converted to an NOEC: NOEC = LOEC/2.
- An LOEC representing an effect more than 20% is not used; an EC_{10} is calculated from the data and is regarded as the NOEC.
- If the percent effect of a LOEC is unknown, then an NOEC cannot be estimated.
- An NOEC may be estimated as MATC/√2.
- An EC_{10} from a long-term test is regarded to be an NOEC.

The EU TGD also describes data reduction procedures for cases in which there are multiple toxicity values for one species (ECB 2003). Values are selected according to which ones reflect realistic European environmental parameters. Also, the database is evaluated to ensure that information will not be lost when averaging procedures are used (e.g., for very sensitive endpoints). After these initial screening steps, results for the most sensitive endpoints are selected. Multiple values for the same endpoint for the same species are evaluated to determine why they are different. If values are determined to be equivalent, then the geometric mean is used. If reasons exist for differences, then results may be grouped according to appropriate factors (e.g., pH ranges). The effects of all possible data exclusions on the final effects assessment must be explored and explained.

With the OECD methodology (1995), if several toxicity results are available for the same species and endpoint, then the geometric mean of values is used. If data are available for the same species, but for different endpoints (i.e., survival vs growth vs reproduction), then only the lowest value is used. The OECD guidelines (1995) require that only chronic NOEC or MATC values be used for statistical extrapolation procedures. If only a chronic LOEC is reported, the NOEC may be calculated as NOEC = LOEC/2, although this conversion is only done if the LOEC corresponds to an effect less than 20%. If the measured effect is greater than 20%, then further toxicity testing is required at lower concentrations. The factor of 2 is representative of the typical interval between test concentrations; if the actual interval is known to be different, then it should be used instead of 2.

For derivation of high reliability TVs, the Australia/New Zealand guidelines (ANZECC and ARMCANZ 2000) require the following: if several NOECs are available for different endpoints for the same species for a particular substance, then the lowest NOEC (i.e., the most sensitive endpoint) is used for criteria derivation; if several NOECs are available for the same effect in the same species for a substance, then the geometric mean of values is used. Acute data, which are used to derive moderate reliability TVs, are reduced using the same approach.

Data reduction may also be necessary for outliers. Although the USEPA (1985) has vague advice regarding data that should be excluded, the Australia/New Zealand guidelines provide clear instructions on how to deal with outlying data (ANZECC and ARMCANZ 2000). First, for excessively high or low data points, original papers are consulted in an attempt to explain the variation (e.g., differences between nominal and measured concentrations, water quality factors, and errors). If data are bimodally distributed, only the lower of the two groups is retained. The use of curve-fitting statistical extrapolation models reduces the need to remove outliers when using this methodology. Data are excluded if they are from unpublished studies or derived from studies with excessively wide test concentration ranges.

Because toxicity data for a given chemical may be available in different forms (i.e., NOEC, LOEC, LC_x/EC_x), for different exposure durations, and for different endpoints, it is necessary to provide some guidance for selecting or standardizing values for use in criteria derivation. Instructions should also be provided on how to reduce multiple data, for a given chemical/species combination, to a single value, as well as for how to manage bimodal distributions and outliers.

7 Criteria Calculation

In this section, we address how criteria values are calculated by different methodologies. Exposure factors that affect toxicity are reviewed, because they may influence how criteria are derived or expressed. Assessment factor and statistical extrapolation methods are described and evaluated and details of criteria calculations are given. Finally, other considerations in criteria derivation are discussed, including chemical mixtures and multiple stressors, bioaccumulation and secondary poisoning, threatened and endangered species (TES), harmonization of criteria across environmental compartments and utilization of data.

7.1 Exposure Considerations

Typically, only effect assessments, and not exposure assessments, are used to derive water quality criteria. However, an exposure component can be included with the effects assessment.

Water quality criteria that adequately protect aquatic life must consider exposures of varying magnitudes, durations and frequencies. A criterion designed to protect against ongoing, chronic toxicant exposure will overprotect if only brief, mild excursions are encountered above the criterion. Similarly, it will be underprotective if brief, but large excursions are encountered. Such situations actually happen in the Sacramento and San Joaquin River basins, where short-term toxicant pulses, coincident with pesticide use or storm events, occur regularly (Bailey et al. 2000; Dileanis et al. 2002, 2003; Domagalski 2000; Dubrovsky et al. 1998; Kratzer et al. 2002; Kuivila et al. 1999; Werner et al. 2000). Many single-species studies have shown that pulse exposures to toxicants can cause significant effects in aquatic organisms that may induce population-level effects (Schulz and Liess 2000; Brown et al. 2002; Ingersoll and Winner 1982; Forbes and Cold 2005; Cold and Forbes 2004; Hodson et al. 1983). However, in mesocosm studies, or in population-level analysis of single-species tests of pulsed pesticide exposures, no long-term effects were found (Heckman and Friberg 2005; Reynaldi and Liess 2005; Pusey et al. 1994). In the latter studies, some period of recovery was required (as rapid or 2–3 week). Presumably, if a community were to receive another pulse exposure before full recovery, effects of the new pulse would be superimposed on those of the first. Thus, it is important to have water quality criteria that are defined in terms of magnitude, duration and frequency, in such a way that monitoring programs can be readily designed to determine exceedances.Two basic approaches are used to address exposure in existing criteria derivation methodologies. First, is to incorporate some combination of magnitude, duration and frequency, in each criterion statement (USEPA 1985; Zabel and Cole 1999; Roux et al. 1996). Second, is to derive only exposure magnitude, and leave decisions on duration and frequency to site-specific management (ANZECC and ARMCANZ 2000; CCME 1999; Lepper 2002; BMU 2001; Irmer et al. 1995; OECD 1995; RIVM 2001; ECB 2003; Bro-Rasmussen et al. 1994; Samsoe-Petersen and Pedersen 1995).

The USEPA criteria (1985) are expressed in terms of magnitude, duration and frequency, with separate acute and chronic criteria. Magnitude is determined by analysis of effects data, but duration and frequency are the same for all toxicants. The allowable exposure durations, expressed as an averaging period of 4 d for chronic toxicity, and 1 hr for acute toxicity, are meant to restrict concentration fluctuations above the criteria in receiving waters. As mentioned, several studies have shown that pulses of high exposure can cause greater effects in single-species toxicity tests, than the same average constant concentration. It follows that minimizing the length of the averaging period will minimize concentration fluctuations during the period. When the USEPA criteria guidelines were developed, there were few studies to support the notion that observed chronic toxicity resulted from toxicant effects on a sensitive life stage over a relatively short period. However, recent USEPA toxicity test guidance (USEPA 2002b) indicates that chronic toxicity may be estimated by sensitive life-stage tests lasting 4–7 d, *in lieu* of full life-cycle tests. Thus, for chronic toxicity, the 4 d averaging period seems reasonable. Four d is long enough to observe the equivalent of chronic toxicity, but minimizes opportunities for concentration fluctuations.

The 1 hr period for acute toxicity is somewhat arbitrary and is based on (1) the fact that it is shorter than the period of a typical acute test; and (2) a nonreferenced opinion that "high concentrations of some materials can cause death in one to three hr" (USEPA 1985). The Technical Support Document for Water Quality-based Toxics Control (TSD; USEPA 1991) indicates that the 1 hr period is derived from data on ammonia toxicity, which implies that it is a very conservative number for toxicants that are not as fast-acting as ammonia. While the supporting data fail to support the 1 hr averaging period, the importance of exposure duration to toxicity is well documented (Newman and Crane 2002). Alternative methods of addressing exposure duration in criteria derivation will be discussed later.

Finally, the frequency of one excursion every 3 yr is intended to allow ecosystem recovery (USEPA 1985). Again, this number appears arbitrary, because it is based on studies that show ecosystems require from 6 week to10 yr to recover from toxicant-induced damage (USEPA 1985). However, the TSD (USEPA 1991) indicates that, although data were lacking in 1985 to relate criteria excursions to ecological effects, the criteria are designed such that a single marginal excursion should cause little-to-no ecological effect. It goes on to argue that if marginal excursions are rare, then high-stress events that require recovery time would be extremely rare, and the 3-yr interval should be very protective.

The UK methodology states criteria in terms of magnitude and duration, but not frequency. An AA concentration is intended to protect ecosystems against long-term exposure, whereas, a MAC is meant to protect against transient concentrations that may cause acute toxicity (Zabel and Cole 1999). Although the AA and MAC are intended to protect against different exposure durations, they generally do not include specific statements regarding duration (such as the 1 hr and 4 d averaging periods stated in USEPA criteria). Defined this way, determinations of whether the AA and MAC are being met or not is dependent on monitoring program design.

In Australia/New Zealand, Canada, EU member nations, and the state of North Carolina, criteria are expressed in terms of magnitude only, and are designed to protect against long-term exposure (ANZECC and ARMCANZ 2000; CCME 1999; BMU 2001; Lepper 2002; Irmer et al. 1995; ECB 2003; Bro-Rasmussen et al. 1994; North Carolina Department of Environment and Natural Resources 2003). In these cases, the values derived are intended to be used by water quality managers to develop enforceable standards (which take into account factors such as use designations and economic considerations), or to trigger further data collection. Thus, allowable frequency and duration of exceedances are part of the management process, rather than the criteria derivation process.

For countries that follow EU guidance, the pesticide criteria reflect values not to be exceeded by the 90th percentile of the levels monitored in water (Lepper 2002); hence, duration and frequency of exceedances depend entirely on monitoring design. An EU Expert Advisory Forum, convened in 2001 and 2002, considered alternatives to analyzing monitoring data for determining WFD goal compliance (Lepper 2002). Possible alternatives included use of annual arithmetic mean, geometric mean, median, 90th percentile, and a maximum never to be exceeded. The Forum concluded that, from a scientific standpoint, either the arithmetic mean or the 90th percentile would be the best measure to determine a reference condition, but the choice between those two was political. In his report, Lepper (2002) proposes that the EU should consider the use of a maximum acceptable concentration value, in addition to quality standards designed to assess annual reference conditions.

The maximum acceptable concentration would be a concentration not to be exceeded any time, and is intended to protect against episodic exposure events.

Within Canada, British Columbia (BC) has its own criteria derivation methodology (Government of British Columbia 1995), which closely resembles that of the CCME (1999), except that the BC guidelines recommend derivation of separate acute and chronic criteria for substances that are known to be acutely toxic. Similarly, South Africa utilizes a modified USEPA methodology (1985), in which final criteria are stated as either acute effect values (AEV) or chronic effect values (CEV; Roux et al. 1996). Thus, both BC and South Africa criteria address the role of exposure duration in toxicity, although they do not address frequency.

Analysis of ecotoxicity data by time to event (TTE) methods allows simultaneous consideration of exposure magnitude and duration when making effects predictions (Newman and Crane 2002). Among other things, the TTE models may be used for (1) estimates of effects over any time period, rather than just at the end of an arbitrary test period; (2) extrapolation from acute-to-chronic exposures; (3) analysis of time-varying exposure (e.g., pulse exposures); and (4) determination of changes in relative risk over time (Crane et al. 2002).

Among current criteria derivation methodologies, only the Australia/New Zealand guidelines (ANZECC and ARMCANZ 2000) allow the use of the TTE methods of Mayer et al. (1994) and Sun et al. (1995) to calculate chronic toxicity values from acute toxicity data. A computer program called ACE (acute-to-chronic estimation; version 2.0) is available from USEPA to do such calculations (USEPA 2003c). Unfortunately, as noted in the Australia/New Zealand guidelines, it is almost impossible to obtain the raw data required to use these models.

The TTE methods are under review in the US and the UK for possible revisions to derivation methodologies. The Water-based Criteria Subcommittee (WCS) of the USEPA is planning to propose that kinetic-based modeling be incorporated into revised guidelines (USEPA 2005). Although the exact model has not been determined, the workgroup, who are considering a model (or models) that will describe the time course of toxicity, will include a toxicant accumulation component. To improve the UK methodology, Whitehouse et al. (2004) has recommend use of survival time modeling, accelerated life testing, and theoretically derived functions that may address the time dependence of toxicity (Dixon and Newman 1991; Newman and Aplin 1992; Newman and McCloskey 1996; Sun et al. 1995). These methods may be used to determine the risk of death within a given time interval, depending on toxicant concentration. Whitehouse et al. (2004) determined that the two-step linear regression method of Mayer et al. (2002) is a relatively easy way to generate LC_0 values (i.e., chronic toxicity values derived from LC_{50} data), which may then be used to construct SSDs for determining hazardous concentrations. Whitehouse et al. (2004) reported that data required for the TTE analysis (i.e., survival at 0, 24, and 48 hr, etc.) is usually collected during standard ecotoxicity tests, but is often not reported (and not obtainable). Thus, this type of analysis would not require new test procedures, but would require new reporting procedures.

The USEPA WCS is considering the use of population models to provide means for criteria to reflect population recovery, after toxic events (USEPA 2005). Such a model was used by the USEPA for deriving dissolved oxygen criteria for the Cape Cod to Cape Hatteras region (USEPA 2000). The WCS notes that population models are complex, and their application may be prohibitively resource intensive. Nonetheless, such models constitute the best way of determining the significance of effects on survival, growth, and reproduction that are typically measured in laboratory toxicity tests. Ultimately, further literature searches for ecosystem recovery studies may be the most practical way to determine appropriate excursion frequencies.

In whatever format criteria are stated, monitoring programs must be designed to include a compliance component. For criteria that are expressed as a single number (ANZECC and ARMCANZ 2000; OECD 1995; CCME 1999; RIVM 2001; Samsoe-Petersen and Pedersen 1995; Bro-Rasmussen 1994; Irmer et al. 1995; Lepper 2002), the risk manager must determine how often and with what frequency a criterion can be exceeded, and then design a monitoring program to assess compliance. For criteria that include duration and frequency components (USEPA 1985; Roux et al. 1996; Zabel and Cole 1999), the risk manager has only to design the monitoring program.

Exclusion of duration and frequency components from criteria statements leaves those factors solely to policy-based decisions. It would be better if these components could be science-based. Although the USEPA approach (1985) of expressing criteria strengthens the science component, the duration and the frequency values, used in acute and chronic criteria statements would benefit from a stronger scientific basis. It is possible that a review of more recent literature could strengthen the duration and the frequency values. The TTE models would give risk managers more science-based information for determining the duration component; similarly, population models and/or good ecosystem recovery studies would help in determining the frequency component.

Aquatic life is exposed to contaminants by two routes: water and food. Water quality criteria derived from single-species laboratory studies are based on water-only exposures, which may considerably underestimate the actual environmental exposure resulting from water and contaminated food sources (Benson et al. 2003). An extreme example is demonstrated in a study of effects of selenium on fish (Lemly 1985). Loss of diversity and reproductive failure occurred in fish communities exposed to selenium at concentrations 10–35 times lower than concentrations causing adverse effects in laboratory studies. Benson et al. (2003) noted that the extent of dose underestimation caused by ignoring food exposure has not been well studied, because the significance of the food pathway has only recently been recognized.

Studies of hydrophobic organic chemicals also show the importance of dietary exposure. A model comparing food and water exposures of PCBs (polychlorinated biphenyls) to lake trout in Lake Michigan, determined that 99% of body burdens came from food exposure. Three-spine sticklebacks accumulated significantly more hexachlorobenzene when feeding on contaminated *Tubifex tubifex,* compared to water-only exposures (Egeler et al. 2001). Other studies have shown that the significance of dietary uptake varies, but the underlying factors that determine food exposure are still unclear. Relationships between log K_{ow} values and dietary uptake of hydrophobic chemicals by fish have been reported, but the relationship is not consistent (Gobas et al. 1988; Qiao et al. 2000). Gobas et al. (1988) found an inverse relationship between log K_{ow} and dietary uptake efficiency, with efficiency decreasing for log K_{ow} values more than 7. In contrast, Qiao et al. (2000) found that gill uptake accounted for 98% of fish body burden for chemicals with log K_{ow} values of 5 or less; whereas, for a chemical with a log K_{ow} value of 7.5, 85% of body burden was from dietary uptake. The Qiao et al. model (2000) determined that food–water concentration ratios were important predictors of the relative uptake by the two routes. For ratios more than 10^7, dietary uptake was predicted to be 100%; at about 10^5, uptake was similar for diet and water; and for ratios less than 10^3, uptake was 100% from water. The relationship between log K_{ow} and exposure route was modeled using environmentally relevant food–water concentration ratios (ranging from 191 to $10^{5.9}$). Fisk et al. (1998) found a significant curvilinear relationship between log K_{ow} and dietary uptake efficiency, with efficiency increasing for log K_{ow} values between 5 and 7, and then declining for values above 7. Other studies have found no clear relationship between log K_{ow} and uptake efficiency. One study (Loonen et al. 1991), suggested a link between level of chlorination of dioxins, and another reported an activated transport mechanism for hydrophobic organic chemicals, with uptake efficiency dependent on molecular weight (Burreau et al. 1997).

Although dietary uptake is an important exposure route for many hydrophobic organic compounds, the theoretical basis for differences in dietary uptake efficiency is not clearly established. For narcotic chemicals (those exhibiting a non-specific mode of action), Traas et al. (2004) have developed a food web model for calculation of environmental quality criteria, based on internal effect concentrations. This model is based on concentrations of contaminants already in organisms, rather than from exposure. Thus, all exposure routes are incorporated. However, the model does not work for chemicals with specific toxic modes of action, a characteristic of most new pesticides.

Until food web or other models are further developed to incorporate multi-pathway (multiple routes) exposures into criteria derivation, it is probably best to continue with water-only assessments. If studies show criteria to be underprotective, and if a substance has a log K_{ow} between 5 and 7, then dietary uptake studies, specific to the compound and species affected, should be performed to determine if exposure has been significantly underestimated. Many modern pesticides tend to be less hydrophobic (even water soluble) rendering the dietary exposure route less important.

When deriving water quality criteria, it is appropriate to also consider whether criteria should be expressed as total chemical or bioavailable chemical, and whether criteria should be adjusted for other factors (e.g., pH, temperature, interactions with other substances) known to affect toxicity. Criteria that utilize toxicity test data are intrinsically based on bioavailable chemicals, and thus incorporate bioavailability. However, many laboratory tests are performed in clean water under controlled water quality conditions. Such tests do not replicate the effects that natural waters have on toxicity. Bioavailability is addressed in virtually all existing methodologies, particularly for metals. Few methods, however, quantitatively address bioavailability of organic chemicals, and the associated influence of pH and temperature on toxicity.

In the UK, EQS may be expressed either as total or dissolved concentrations, though expert judgment determines how it is done. The Canadian protocol provides no specific methodology to account for water quality factors that may affect toxicity (CCME 1999). Pawlisz et al. (1998) derived Canadian water quality guidelines for the pyrethroid insecticide deltamethrin [(*S*)-cyano(3-phenoxyphenyl)methyl (1*R*,3*R*)-3-(2,2-dibromoethenyl)-2,2-dimethylcyclopropanecarboxylate], and although they acknowledged that deltamethrin toxicity to insects is temperature-dependent, they did not address that issue in deriving the criteria. The EC's TGD on Risk Assessment (ECB 2003) discusses effects of pH on bioavailability and toxicity of ionizable organic chemicals in Appendix XI. It indicates that toxicity tests ought to be conducted at pH levels above and below the pK_a for the test substance. However, because this is rarely done (because toxicity tests must be conducted in narrowly prescribed pH ranges) effects of pH on toxicity can only be qualitatively discussed as part of a risk assessment.

The Australia/New Zealand guidelines acknowledge that suspended solids, dissolved organic matter, and total organic carbon levels in water may affect bioavailability, and thus toxicity, of organic compounds. However, the guideline authors did not believe that such solid–toxicant interactions are understood well enough to allow specific quantitative guidance for national criteria setting. Guidance is only given for case-by-case, site-specific evaluation of bioavailability. If quantitative relationships exist between toxicity and a parameter affecting that toxicity, such as pH or temperature, then factors may be applied to calculate a site-specific target value. When a generally applicable quantitative relationship is absent, the use of direct toxicity assessment (DTA) using local waters and local conditions is recommended (ANZECC and ARMCANZ 2000).

A few methodologies offer specific guidance on how to express criteria as either total or dissolved concentrations. In Germany, if a substance has a suspended particulate matter–water partition coefficient greater than 1000 L/kg, the target is expressed as the level in suspended particulate matter, and is calculated as follows (Irmer 1995, adapted from LAWA 1997):

$$QT_{SPM}(\mu g/kg) = QT_{water}(\mu g/L) \bullet \frac{K}{10^{-6} \bullet K \bullet 25(mg/L) + 1} \tag{2.1}$$

where

QT_{SPM} = quality target in suspended particulate matter (μg/kg);
QT_{water} = quality target in water, total (μg/L)
K= partition coefficient (L/kg)
25 = default concentration of suspended particulate matter (mg/L)
10^{-6} = conversion factor (kg/mg)

In the Netherlands, ERLs (MPC and NC) for water are reported, both for dissolved and total concentrations, based on a standard amount of suspended matter (30 mg/L). The total concentration is calculated as follows (RIVM 2001):

$$\mathrm{MPC}_{\mathrm{water\_total}} = \mathrm{MPC}_{\mathrm{water\_dissolved}}(1 + K_{\mathrm{ppm}} \times 0.001 \times 0.03) \tag{2.2}$$

where
0.001 = conversion constant (g/kg)
0.03 = content of suspended matter (g/L)
K_{ppm} = partition coefficient for suspended matter/water
and

$$K_{\mathrm{ppm}} = K_{\mathrm{oc}} \times f_{\mathrm{oc}} \tag{2.3}$$

where
K_{ppm} = partition coefficient for standard suspended matter (L/kg)
K_{oc} = organic carbon-normalized partition coefficient (L/kg)
f_{oc} = fraction organic carbon (standardized at 11.72%)
and,

$$NC_{\mathrm{water\_total}} = \frac{MPC_{\mathrm{water\_total}}}{100} \tag{2.4}$$

where 100 is a safety factor to account for mixture effects.

Both the German and Dutch methods depend on an assumption of a standard concentration and composition of solids for determining solid–water partitioning. Unfortunately, partition coefficients are highly dependent on the composition of the solids and on the nature of the contaminant (Schwarzenbach 1993). Solid–water partition coefficients can be underestimated if colloids are not removed from the solution phase (Wu and Laird 2004). If partition coefficients are not specific to sediments in a given sample, calculations of dissolved versus bound pesticides may produce erroneous results. For example, Wu and Laird (2004) determined that partition coefficients for chlorpyrifos [*O,O*-diethyl *O*-(3,5,6-trichloro-2-pyridinyl) phosphorothioate] in aqueous mixtures of six different smectites ranged from 45 L/kg to 6,846 L/kg. Burgess et al. (2005) found partition coefficients for nonylphenol that ranged from 21.3 for cellulose to 9,770 for humic acid, indicating that even if values are normalized to organic carbon content they may not produce applicable partition coefficients. Selecting a single value from such a wide range to represent the partitioning behavior for solids of all compositions makes little sense.

The Dutch methodology recommends normalizing ERLs to a specific pH, or to base the ERLs on the relevant chemical species, for chemicals whose speciation, and thus bioavailability and/or toxicity, depend on pH (RIVM 2001). This adjustment would apply to weak organic acids, such as phenols. Degradation of compounds and metabolite formation are also considered in this methodology. Compounds with half-lives less than 4 hr must have the MPC derived from stable degradates or metabolites.

The USEPA (1985) provides detailed instructions for determining acute and chronic criteria where toxicity to two or more species is related to a water quality characteristic (hardness, pH, temperature, etc.). This method is not only regularly applied to metals criteria but also applies to pesticides that have pH- or temperature-dependent toxicity. The key is that a demonstrable quantitative relationship must exist between toxicity and the water quality parameter. In such cases, criteria are expressed as mathematical formulae that describe that relationship. The USEPA "Guidelines for Deriving Numerical Aquatic Site-Specific Water Quality Criteria by Modifying National Criteria" describes the water-effect ratio (WER) technique, designed to account for differences in bioavailability, that is dependent on chemical–physical characteristics of site water (USEPA 1984a), as follows:

$$\text{Water\_Effect\_Ratio} = \frac{\text{Site\_Water\_LC}_{50}}{\text{Laboratory\_Water\_LC}_{50}} \tag{2.5}$$

The site-specific maximum concentration is then equal to national maximum concentration multiplied by the WER.

The site water to be used in WER determination is to be collected under typical conditions (i.e., not during floods or storms). However, because pesticide loadings to surface waters typically result from storms or agricultural runoff, and because suspended solid content is higher than normal during runoff events, it is best to express criteria in terms that reflect the covariance of pesticides and suspended solids at the time a sample is taken. The simplest method would be to derive criteria based on dissolved concentrations (as is typically done), and then to use solids data, together with measurements of total concentrations and partition coefficients, to determine compliance. This could be achieved using the following equation, which is given in RIVM (2001), for converting total concentrations to dissolved concentrations:

$$C_{\text{dissolved}} = \frac{C_{\text{total}}}{1+(KS)} \tag{2.6}$$

where

$C_{\text{dissolved}}$ = concentration of chemical in dissolved phase

C_{total} = total concentration of chemical in water

K= solid–water partition coefficient (L/kg); may be expressed as K_{oc}/f_{oc}

S= concentration of sediment in water (kg/L)

The resulting dissolved concentration is then compared to the water quality criterion, to determine compliance.

In the Central Valley of California, levels of suspended solids vary greatly. The US Geological Survey reports levels ranging from 1 mg/L to 330 mg/L, in samples from various streams in the Sacramento River Basin, and from 1 mg/L to 5280 mg/L, in the San Joaquin River Basin (USGS 2005a, b). For pesticides with high sediment–water partition coefficients, bioavailability may vary considerably with solids levels and, ideally, this factor should be considered in deriving water quality criteria.

In addition to bioavailability, effects of other water quality factors should be considered in deriving criteria. For organic chemicals, these factors include pH and temperature. As described in USEPA (1985), if data are available to establish quantitative relationships between water quality characteristics and toxicity, then criteria should be expressed as equations reflecting that relationship.

7.2 Basic Methodologies

Two basic criteria derivation methodologies are in use, or proposed for use, throughout the world. The aim of both methods is to extrapolate values from available toxicity data to ones that will protect the environment. The first of these two methods is the AF method. It involves multiplying the lowest value of a set of toxicity data by a factor to arrive at a criterion. The second method is the statistical extrapolation method. It involves the use of one of several similar SSD techniques to determine the criterion. Some countries exclusively use one of these methods, and others use a combination of the methods, depending on data availability. In a 1993 review of the statistical procedure of Van Straalen and Denneman (1989), and the AF method used by USEPA (1984b; now in Nabholz 1991), Calabrese and Baldwin (1993) concluded that these two methods produced the same results and no strong argument exists for selecting one method over the other.

One advantage of the SSD approach over the AF method is that it provides for deriving a criterion with a known level of confidence. Unfortunately, the SSD method utilized by the USEPA (1985) does not allow the inclusions of confidence levels. However, other SSD methods do provide for quantification of confidence levels (RIVM 2001: ANZECC and ARMCANZ 2000).

France, Germany, Spain, the UK, and Canada utilize only the AF method for derivation of water quality criteria (Lepper 2002; BMU 2001; Zabel and Cole 1999; CCME 1999). Others, including Australia/New Zealand, the Netherlands, USEPA, the EU, Denmark, and OECD utilize a combination of the SSD and AF methods (ANZECC and ARMCANZ 2000; RIVM 2001; USEPA 1985; ECB 2003; Bro-Rasmussen et al. 1994; Samsoe-Petersen and Pedersen 1995; OECD 1995).

In France, Spain, Germany, and the UK criteria are derived by multiplying (or dividing) the lowest toxicity value from a minimum data set by a factor. One criterion is derived that is supposed to protect against long-term exposures (Lepper 2002; Irmer et al. 1995; BMU 2001). In France, AFs of 1–1000 are applied to single-species toxicity values. For derivation of low-level criteria, acute data may be used with an AF of 1, but high-level criteria are derived by applying an AF of 10 to chronic NOEC data, or 1000 to acute data (Lepper 2002). In Spain, data corresponding to the most sensitive organism are used in criteria derivation. LC_{50}/EC_{50} values are multiplied by a safety factor of 0.01 and chronic NOEC values by a factor of 0.1. Further safety factors are applied to account for lack of relevant species, persistence or bioaccumulative potential and genotoxic potential (Lepper 2002).

In the UK, the lowest relevant and reliable adverse effect concentration in the data set is multiplied by a safety factor. An MAC, to protect from acute toxicity, is derived from acute data, with a factor of 2–10 applied to the lowest available acute toxicity value. An AA concentration to protect from chronic toxicity may be derived from either acute or chronic data, or from acceptable field data, with application of appropriate factors (from 1 to 100) to the lowest available toxicity value (Zabel and Cole 1999).

The Canadian methodology (CCME 1999) uses chronic LOEC values to derive criteria. If there is an adequate data set, then the lowest LOEC is divided by a factor of 10. If only acute data are available, then the lowest LC_{50}/EC_{50} value is divided by an ACR, if one is available. The resulting estimated chronic value is then divided by 10 to derive the criterion. If no ACR is available, then the criterion is derived directly from the lowest LC_{50}/EC_{50} by dividing it by either 20 (for nonpersistent chemicals) or 100 (for persistent chemicals).

The Netherlands methodology (RIVM 2001) utilizes the AF method for derivation of MPC and SRC_{ECO} values, through a process called "preliminary effect assessment." This is not the preferred derivation method, and is used only where results of four chronic toxicity studies, from four different taxonomic groups, are not available. Assessment factors range from 1 to 1000, and are applied according to the amounts and kinds of data available. Similarly, the OECD recommends use of an AF method for limited data sets (OECD 1995), with factors ranging from 1 to 1000, depending on available data. A factor of 10 is applied to the lowest NOEC or QSAR estimate of chronic toxicity, from a data set that includes at least algae, crustaceans, and fish. If only acute data or QSAR estimates of acute data are available, then a factor of 100 is applied, if the data set includes algae, crustaceans, and fish; a factor of 1000 is applied only if one or two species are represented.

Although the USEPA (1985) does not derive criteria when data sets are inadequate, the state of North Carolina and the Great Lakes region utilize the AF method to derive criteria when data are lacking. In addition, for derivation of criteria in California, Lillebo et al. (1988) developed an AF method that uses LOEC values. For pesticides, this method involves finding the geometric mean of the three lowest LOEC values from acceptable tests, and multiplying them by a factor of 0.1. This criterion is intended to protect all species in an ecosystem from the effects of long-term exposure. In the state of North Carolina, if adequate data are not available for derivation of a FAV, using the USEPA methodology (1985), then a factor of 3 is applied to the lowest available LC_{50} value to render an acceptable acute value. North Carolina uses chronic toxicity values to set aquatic life standards. In the absence of a chronic value, a measured ACR may be applied to an acute value. If no ACR is available, then the acute value is divided by 100 (for $t_{1/2} > 96$ hr), or 20 (for $t_{1/2} < 96$ hr; North Carolina Department of Environment and Natural Resources 2003).

For derivation of Tier I aquatic life values, the Great Lakes methodology (USEPA 2003a) follows the USEPA guidelines (1985). However, when not enough data are available for derivation of Tier I values, Tier II values are derived using an AF method. Secondary acute values (SAVs) are derived by dividing the lowest available GMAV by a factor ranging from 4.3 (if seven GMAVs are available) to 21.9 (if only one GMAV is available). The secondary maximum concentration (SMC) is the SAV divided by 2. The secondary chronic value (SCV) is derived in one of three ways: (1) the FAV (from a Tier I procedure) is divided by a secondary ACR(derivation is described below); (2) the SAV is divided by the final ACR (from Tier I); or (3) the SAV is divided by the SACR. The secondary chronic concentration (SCC) is equal to the lower of the SCV or the Final Plant Value (FPV).

In practice, all of the current Australia/New Zealand TVs that were derived from single-species toxicity tests were calculated by the SSD method, but the ANZECC and ARMCANZ guidelines (2000) include an AF method where data are lacking.

Some of the TVs were derived by applying a factor of 10 to the lowest of at least three acceptable multiple species tests. To derive moderate reliability TVs, when only acute data for more than five species are available, a factor of 10 is applied before applying the ACR. No justification for choosing a factor of 10 is given. Low reliability TVs are derived by applying factors that range from 20 to 1000; larger factors are applied when data sets are smaller or contain more acute than chronic data.

The Danish methodology utilizes the EU AFs for its AF method (Samsoe-Petersen and Pedersen 1995); however, if the SSD method (that of Wagner and Løkke 1991) is used, application of default ACRs to derive NOECs is not allowed.

The South African methodology (Roux et al. 1996) follows that of the USEPA (1985) very closely, except that the FAV is divided by one of several safety factors (rather than 2) to derive the AEV. The FCV is calculated as in the USEPA guidance (1985), but again, safety factors ranging from 1 to 1000 are applied to derive the CEV. If no chronic data and no ACRs are available, a CEV is derived by multiplying the FAV or the FPV by 1000. The FPV is the lowest result from a 96 hr algae test or from a chronic test with a vascular plant.

AF, safety factor, application factor, ACR, and margin of safety are all terms that refer to a value that is used as a multiplier for experimentally determined toxicity values; the values described by these terms are all designed to account for the uncertainty of using that experimentally derived number to predict real-world outcomes. Chapman et al. (1998) reviewed the use of safety factors in ecological risk assessment. They point out that, despite a lack of supporting data, standardized factors of 0.1, 0.05, and 0.01 are used throughout the world in various regulatory programs (often expressed inversely, i.e., 10, 20, and 100, for use as divisors). They also note that safety factors are applied as policy to assure protection, rather than being based on empirical science.

According to Irmer et al. (1995), factors used in derivation of German water quality criteria were based on internationally accepted practices until 1992, when political motivations limited factors to a total value of 0.01. The factor of 0.1 is applied to extrapolate from lab, single-species tests, to field conditions; and a further factor of 0.1 may be applied to protect against various uncertainties. Yet an additional acute-to-chronic factor of 0.1 may be applied, when chronic data are not available. Irmer et al. (1995) argue that limiting the total applied safety factor to 0.01 results in weak water quality targets, because ACRs as high as 1000 are not uncommon.

Factors used in preliminary effect assessment, in the Netherlands, are derived from two sources. First, is the TGD for derivation of the PNEC (ECB 2003). Second, is a USEPA (1984b) document cited by Van De Meent et al. (1990). A more recent version of the same USEPA procedure is now available (Nabholz 1991). The TGD (ECB 2003) factors address the uncertainty associated with intra- and interlaboratory variation in toxicity data, intra- and interspecies variations, short- to long-term toxicity extrapolation and laboratory to field extrapolation (which includes mixture effects). For each of these extrapolations, a factor (divisor) of between 1 and 10 is applied, and if multiple extrapolations are required, then the factor can be as high as 1000. The USEPA document (Nabholz 1991) recommends factors ranging from 1 to 1000, depending on the amount and type of data available. Factor size also depends on uncertainties arising from making extrapolations from acute-to-chronic, laboratory to field, and from small (i.e., $n = 1$) acute data sets. If data are available, an acute-to-chronic factor may be calculated, rather than using a default value of 10.

In the UK, factors are used to deal with uncertainty arising from extrapolations such as one species to another, short to long exposure times, acute-to-chronic effects, chronic to ecosystem effects, and effects in one ecosystem to those in another (Zabel and Cole 1999). Factor size depends on the quantity of data and whether data are available for sensitive species. An additional factor may be applied if a substance is bioaccumulative (usually substances with MW < 700 and BCF > 100 or log $K_{ow} > 3$). Factors range from 1 to 1000, and the degree to which they are applied relies heavily on expert judgment.

According to the OECD guidelines (1995), AFs are empirically derived; that is, they have no theoretical basis. In this methodology, a factor of 10 is applied for each of three possible extrapolation steps: (1) laboratory-derived NOEC to those in the field; (2) short to long exposure times; and (3) acute-to-chronic effects. Alternatively, the OECD (1995) provides for use of AFs presented in the EU risk assessment methodology (ECB 2003).

ACRs are used in the USEPA methodology (1985) to derive chronic criteria when chronic data are lacking. ACRs are calculated from chronic data, for which at least one corresponding acute value is available (from the same study, or from a different study using the same dilution water). Species mean ACRs are calculated as the geometric mean of all available ACRs, for that species. To calculate the final ACR, one of four methods is used: (1) for materials for which the species ACRs covary with the SMAV, the ACR is calculated using only species whose SMAVs are close to the FAV ("close" is not defined); (2) if there is no covariance, and all of the ACRs for a set of species are within a factor of 10, then the final ACR is calculated as the geometric mean of all species ACRs, including both freshwater and saltwater species; (3) for acute tests with shellfish embryos and larvae, a final ACR of 2.0 is used, which makes the FCV equal to the criterion maximum concentration (CMC); and (4) for species with mean ACRs less than 2.0, a final ACR of 2.0 is used to provide for possible acclimation of test species to the toxicant. If a final ACR cannot be determined by any of these methods, then it is likely that neither the final ACR nor the FCV can be calculated.

Factors used in deriving SAVs in the Great Lakes guidance range from 4.3 to 21.9, depending on how many of the minimum Tier I data requirements are met. For example, if seven toxicity values from different families are available, then the factor is 4.3, but if only one value is available, then the factor is 21.9. According to Pepin (2005, personal communication), these factors are based on a USEPA study by Host et al. (1995), which presents several methods for deriving factors to use for data sets that are smaller than the minimum eight values. An SACR is derived by using any available measured ACRs plus enough assumed ACRs of 18 (default value adopted for the Great Lakes methodology) to give a total of 3 ACRs.

For example, if no measured ACRs are available, then three assumed ACRs (of 18) are used. If two measured values are available, then just one assumed value is used. The geometric mean of the three values is then used as the SACR, which is used to calculate the SCV, just as the ACR is used to calculate and FCV. For the AF method, employed by the state of North Carolina, a factor of 3 is applied to the lowest available LC_{50} value. Factors ranging from 20 to 100 are used as default ACRs. No justification for these factors is given (North Carolina Department of Environment and Natural Resources 2003). Lillebo et al. (1988) use an additive toxicity model to derive the suggested factor of 0.1; this factor is applied to the geometric mean of the three lowest LOECs among acceptable studies. Because this value was derived from metals effects data, the applicability to pesticides may or may not be valid.

The factor of 10, used in deriving TVs from multispecies data in Australia/New Zealand, is to account for variations in mesocosm types, and for the fact that more sensitive species may not have been in the test systems (ANZECC and ARMCANZ 2000). No particular justification is given for factors used to derive moderate and low reliability TVs; however, they are similar to those provided in the OECD guidelines (OECD 1995), on which much of the Australia/New Zealand methodology is based. Acute-to-chronic conversions are accomplished, in the Australia/New Zealand guidelines, in one of three ways: a chemical-specific ACR is applied; an LC_0 is calculated according to Mayer et al. (1994) and Sun et al. (1995); or a default ACR of 10 or more is applied. The chemical-specific ACR is the ratio of an acute EC_{50} to a chronic NOEC. If multiple ACRs are available, the geometric mean of all ACRs for all species is used for derivation of criteria by the SSD method, while the ACR for the most sensitive organism is used for the AF method.

The factors used in the EU guidance (Bro-Rasmussen et al. 1994) range from 10 (to account for experimental variability), to 100 (to account for lack of NOEC data), to 1000 (to account for lack of NOEC and LC_{50} data). Although not specifically stated, the discussion in Bro-Rasmussen et al. (1994) suggests that the final EU factor could be adjusted, if judged necessary because of bioaccumulative potential, persistence, carcinogenicity, mutagenicity, or another concern. The French and the Spanish methodologies utilize the EU factors, although Spain includes the possibility of a factor as high as 100,000, if only acute data are available, if ecotoxicity data for relevant species is lacking, a substance is persistent or bioaccumulative, and a substance has genotoxic potential (Lepper 2002). The EU risk assessment TGD (ECB 2003) uses AFs ranging from 1 to 1000; factor size depends to a large extent on professional judgment. The factors are intended to account for variability of laboratory toxicity data, variability within and between species, short to long-term exposure extrapolation, and laboratory to field extrapolation (which includes effects of mixtures). If more toxicity data are available for species of different trophic levels, different taxonomic groups, and different lifestyles, smaller factors are applied. If only one acute value is available from each of three trophic levels, a factor of 1000 is applied. If only a single chronic NOEC is available from either a fish or daphnid test, a factor of 100 is applied. If two long-term NOECs from two different trophic levels exist, the factor is 50. A factor of 10 is used if chronic

NOECs are available from at least three species representing three different trophic levels. Factors of 1–5 are applied to results of SSD extrapolations. For field or model ecosystem data, the size of the factor is determined on a case-by-case basis.

Acute and chronic safety factors used in the South African methodology (Roux et al. 1996) are intended to compensate for missing information. They are applied to compensate for insufficient data being available to assess inter- and intraspecies variability, or if chronic data are absent. Acute safety factors range from 1 to 100 and depend on completeness of the data set. For example, if the minimum data set is available, and includes results from more than one test in at least three taxonomic groups, then a factor of 1 is applied. At the other extreme, if only one acute result is available, a factor of 100 is applied to the FAV. The chronic safety factors range from 1 to100 and also depend on completeness of the data set. For example, if the chronic database contains ACRs or chronic exposure data for at least one species from three different taxa (including at least one fish), then the factor is 1. If only acute data are available, then a factor of 1000 is applied to the FAV or the FPV to arrive at the CEV. In the South African methodology, ACRs are derived by dividing the geometric mean of available acute values by the geometric mean of chronic values, where acute and chronic values were obtained in the same test, or in tests run in similar water dilutions. The same concerns discussed for the USEPA methodology, regarding covariance of SMAVs and ACRs, and for intraspecies ACR variability, apply in the South African methodology.

Canadian factors range from 10 to 100, but the total factor applied could be higher if, for example, a measured ACR is higher than 10. A factor of 10 is applied to chronic data to account for variability in species sensitivity, extrapolation from laboratory to field, and differences in test endpoints. Higher level factors are applied to acute data when no chronic data are available, and are used to extrapolate from acute-to-chronic exposures, or to derive criteria directly, as described earlier.

There is no theoretical basis for any of the AFs used by the various criteria derivation methodologies. They are all empirically derived numbers. The origin of generic factors of 10, for each step of uncertainty, is not clear; those methodologies that explain the reason for the selection of a value of 10 simply state that it is widely accepted. Measured ACRs seem to have a firmer basis in empirical evidence, but they are usually derived for a particular chemical and for a particular species and are then applied to other species or groups of species, which may lead to further uncertainty in final criteria values.

Different default ACRs are used in different methodologies, when no measured ACR is available. The Great Lakes guidance uses a value of 18 (USEPA 2003a), Canada uses either 2 or 10 (CCME 1999), the OECD, the USEPA's OPPT, and Australia/New Zealand use 10 (OECD 1995; Nabholz 1991; ANZECC and ARMCANZ 2000). Kenaga (1982) reports that ACRs were less than 25 for 86% of 84 chemicals tested. However, for pesticides, 70% of ACRs were more than 25, with the largest at 18,100 for propanil. The large percentage of chemicals with ACRs less than 25 resulted from the fact that 93% of industrial organic chemicals fell into that category. Based on Kenaga's results, the USEPA "Guidelines for Deriving Ambient Aquatic Advisory Concentrations" (USEPA 1986) use a default ACR of 25 for calculation of advisory values, but only for low molecular weight nonionizable organic chemicals. There is no evidence that default ACR values are appropriate for pesticides, in general.

All of the AF methodologies, with the exception of the Great Lakes Tier II procedure (USEPA 2003a), consider data for aquatic animals and plants together in criteria derivation. The criterion is based on the most sensitive species, regardless of such factors as taxon or toxicant mode of action. However, separate freshwater and saltwater criteria are typically derived. There is an issue of whether taxa should be pooled or not and it is of higher concern in SSD extrapolation procedures. This will be discussed, in depth, later in this section.

Assessment factors are recognized to be a conservative approach for dealing with uncertainty when risks posed by chemicals are being assessed (Chapman et al. 1998). Chapman et al. (1998) also note that application of empirically based factors to toxicity data neither quantifies uncertainty nor reduces the probability of underestimating risk. Similarly, the use of AFs also greatly increases the possibility of overestimating risk. Chapman et al. (1998) are very concerned that AFs are typically applied generically, when they should be derived and used when specific factors require it, such as the scale, frequency, and severity of potential environmental insults, or the steepness of a toxicant's dose-response curve. In their conclusion, Chapman et al. (1998) suggest the following principles be applied for using safety factors: (1) data supersede extrapolation; that is, if data are available, they should be used; (2) extrapolation requires context; use of safety factors should be based on existing scientific knowledge; (3) extrapolation is not fact; estimates of effect levels obtained using safety factors should only be used as screening values, not as threshold values; (4) extrapolation is uncertain; safety factors should encompass a range rather than being a single value; (5) all substances are not the same; safety factors should be scaled relative to different substances, potential exposures, and nature of effects; and (6) unnecessary overprotection is not useful; safety factors for individual extrapolation steps should not exceed 10, and may be much lower.

Specifically addressing ACRs, Chapman et al. (1998) cited studies showing that measured ACRs can vary from 1 to 20,000. In view of this, it is unreasonable to apply a generic factor (10 or another number) across species and across substances, as is often done in criteria derivation if no chronic data are available. The reality remains, though, that adequate chronic data are generally not available and some means of extrapolation is needed. If an ACR is developed according to the principles for the use of safety factors (described above), then it will be derived in the context of the best scientific understanding of the substance and of the species under consideration, and should be a better predictor of chronic toxicity than a generic factor would be.

One possibility for reducing the need to use ACRs is found in the work of Duboudin et al. (2004) who have proposed a novel way of directly using acute toxicity data to determine a chronic HC_5 value. By using an acute-to-chronic transformation procedure, derived from comparisons of acute and chronic SSDs within species categories, an acute data set is transformed into a chronic data set, which is then used to determine the HC_5 value.

An alternative to the use of AFs is employing statistical extrapolation methods; such methods use single-species toxicity data to make ecosystem predictions. These methods, discussed next, still rely on acute-to-chronic extrapolation.

According to Posthuma et al. (2002a), many ecologists and ecotoxicologists have independently developed methods for predicting ecosystem-level effects using single-species toxicity data. The USEPA was first to use the SSD method to derive water quality criteria (Suter 2002). In Europe, Kooijman (1987) developed the concept of deriving a hazard concentration for sensitive species (HCS; Van Straalen and Van Leeuwen 2002). According to Van Straalen and Van Leeuwen (2002) Kooijman's idea was soon refined and modified by Van Straalen and Denneman (1989), Wagner and Løkke (1991), and Aldenberg and Slob (1993), and Aldenberg and Jaworska (2000). The most recent version of these SSD techniques appears in the ANZECC and ARMCANZ criteria derivation methodology (2000). The main difference among these methodologies is in selection of the shape of the distribution that is used for extrapolations. Other differences include kinds and quantity of data used in the extrapolation procedures, level of confidence associated with derived criteria, and how data are aggregated to construct the distribution. One area in which there is much agreement among methodologies is in the selection of the 5th percentile as the cutoff for prediction of no-effect concentrations. Figure 1 provides a general illustration of the SSD technique, and will be referred to in the following discussion.

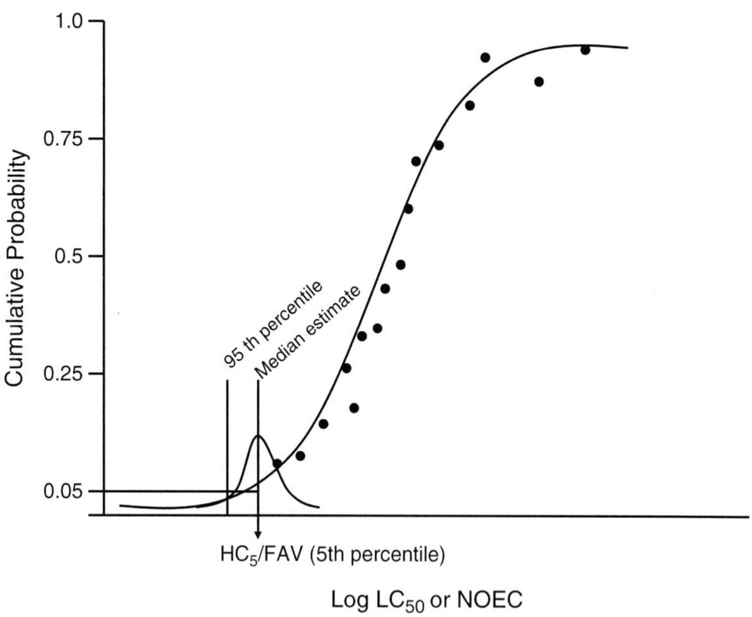

Fig. 1 Generic illustration of the species sensitivity distribution (*SSD*) technique; HC_5 hazard concentration potentially harmful to 5% of species, *FAV* final acute value, *NOEC* no observable effect concentration

The first step in the SSD methodology is to plot data in a cumulative frequency distribution. One approach for doing this is to assume that those data are a random sample of all species, and that if all species were sampled they could be described by one distribution. The USEPA (1985) assumes a log-triangular distribution, while the Netherlands methodology utilizes a log-normal distribution (Aldenberg and Jaworska 2000). The USEPA Office of Pesticide Programs utilizes a log-normal regression method for ecological risk assessment (Fisher and Burton 2003). Any SSD method that utilizes all available data may be used either to determine the percentage of species that could potentially be harmed by an expected environmental concentration, or, conversely, to determine an environmental concentration that will protect some percentage of species. The OECD methodology (1995) offers a choice of the log-normal distribution method of Wagner and Løkke (1991), the log-logistic distribution method of Aldenberg and Slob (1993), or the triangular distribution of USEPA (1985), depending upon which distribution best fits the available data. Figure 2 illustrates the character of the log-normal, log-logistic, and log-triangular distributions.

The OECD method (1995) cites two advantages of the USEPA method (1985). First, because it uses a subset of the lowest available values, it is not affected by deviations of the highest values from the assumed distribution. Moreover, data reported as "greater than" may be used, which is not possible with other methods (Erickson and Stephan 1988). Okkerman et al. (1991) criticize the USEPA's selection of the triangular distribution because it implies a toxicity threshold and possibility of a 100% protection level, and it only uses four (usually the lowest four) data points to calculate a criterion. The authors of the Australia/New Zealand guidelines

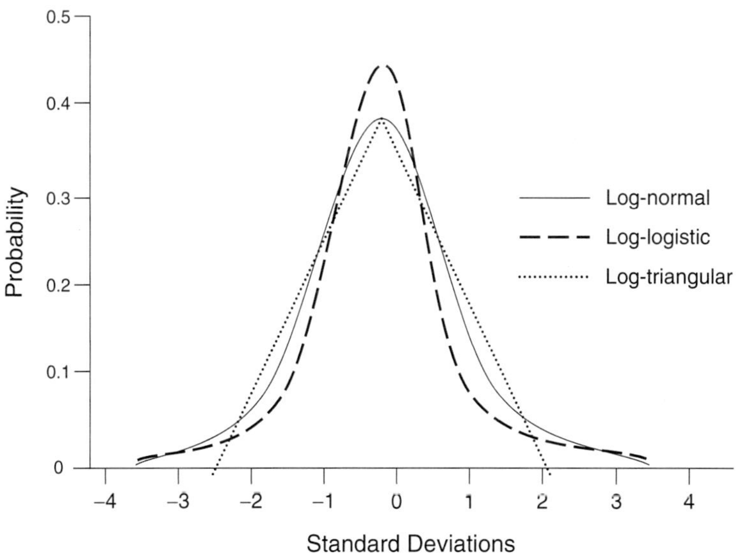

Fig. 2 Comparison of log-normal, log-logistic, and log-triangular distributions

(ANZECC and ARMCANZ 2000) did not adopt the USEPA SSD approach (1985), because its data requirements are too stringent, there is no biological basis for selecting a triangular distribution, not all of the data are used, and it assumes that a threshold toxicity value exists. In defense of the USEPA approach (1985), Erickson and Stephan (1988) argue that, because the entire data set is used in setting percentile ranks and cumulative probabilities, calculation of the FAV using the four data points nearest the 5th percentile does not constitute "not using all the data." Their interpretation is that, those four data points are used as a means of giving more weight to toxicity values nearest 5th percentile. This weighting, however, leads to other problems, which are discussed below.

To sidestep the issue of choosing an appropriate distribution, several researchers believe it best to make no assumptions about distribution shapes, and to use nonparametric methods to estimate community or ecosystem effects, where they are based on single-species toxicity tests (Jagoe and Newman 1997; Van Der Hoeven 2001; Grist et al. 2002). Grist et al. (2002) found that HC_5 values, determined by parametric versus nonparametric methods, were significantly different from each other. Wheeler et al. (2002) suggest that, to get the best HC_5 (or, generally, HC_p) estimate, data should be analyzed by four different SSD methods (two parametric and two nonparametric), selecting the one that gives the best fit. While bootstrapping, techniques offer a solution to the distribution problem, they are very data intensive, and will not work for many small data sets available for criteria derivation.

Arguments for one or the other distribution, or for making no distributional assumptions, are based on which distributions are easier to work with, or which ones better quell the criticism that SSDs are not valid, because data usually do not fit the assumed distribution. Ultimately, all methods currently in use appear to derive protective criteria. In the Netherlands, the log-normal distribution was selected over a log-log distribution (Aldenberg and Slob 1993), because the distributions are not so different, results obtained are not different, and the normal distribution provides powerful statistical tools (RIVM 2001). Similarly, the OECD (1995) concludes that the log-normal, log-logistic, and triangular distribution methods give very similar results.

The ANZECC and ARMCANZ guidelines (2000) take the data-fitting idea a step further in a modification of the Dutch approach. In the Australia/New Zealand methodology data are fitted to one of a family of Burr distributions (Burr 1942; Shao 2000), and then the HC_5 extrapolation is performed using the best fit. This approach allows for derivation of high and moderate reliability TVs from data that would have precluded using log-normal or log-logistic distributions. Noting that the Dutch (MHSPE 1994; RIVM 2001) and Danish (Samsoe-Petersen and Pedersen 1995) SSD methodologies give very similar results, and differ only in the selection of either a log-logistic (Dutch; Aldenberg and Slob 1993) or log-normal (Danish; Wagner and Løkke 1991) distribution, the Australia/New Zealand guidelines chose to start with the Dutch approach, because it had been more extensively evaluated and was easier to use. Advantages to the Dutch approach include that it uses the full range of available data, and a water manager can choose a level of protection and a level of uncertainty associated with a guideline value.

To use an SSD method for criteria derivation requires a policy decision about what percentile of the distribution translates to a protective concentration. The SSD methodologies reviewed here, for example, derive criteria using the 5th percentile of the distribution (Fig. 1). Some methodologies refer to this concentration as an HC_5 (hazardous concentration affecting 5% of species). This is often interpreted to mean that species lying above the 5th percentile in the distribution will be protected, and those species lying below this point will be harmed (e.g., Lillebo et al. 1988, in reference to the USEPA methodology 1985). However, Van Straalen and Van Leeuwen (2002) provide a more accurate interpretation. They assert that the HC_5 does not mean that 5% of species will be harmed. Rather, the HC_5 approach is one for deriving a PNEC, and although the choice of the 5th percentile is purely pragmatic, it has been validated by field studies. Solomon et al. (2001), believe that any percentile is appropriate if it can be validated against knowledge and understanding of ecosystem structure and function. Following is a discussion of percentile cutoff values used by selected methodologies, why they were chosen, and whether they have been validated.

The USEPA rationale for choosing the 5th percentile is that criteria values derived using the 10th or 1st percentiles were thought to be too high and too low, respectively; the 5th percentile was selected because it falls between 1 and 10 (Stephan 1985). Nonetheless, there is good agreement between USEPA criteria and no-effect concentrations determined in experimental stream studies (USEPA 1991). The Dutch guidelines (RIVM 2001) use the 5th percentile for derivation of MPC values, and the 50th percentile for calculation of the SRC_{ECO}. The reasons for these choices are not given, but the 5th percentile has been validated against field NOECs, in studies by Emans et al. (1993) and Okkerman et al. (1993). The Australia/New Zealand guidelines (ANZECC and ARMCANZ 2000) consider the question more rigorously, but still arrive at the 5th percentile level, because it works well in the Dutch guidelines (RIVM 2001) and it gives TVs that agree with NOEC values from multispecies tests. The reason for not regularly using a lower percentile is that uncertainty is very high in the extreme tail of the distribution, and such uncertainty may contribute more to the derived TV than does the data. However, the Australia/New Zealand guidelines do use the 1st percentile as a default value for high conservation ecosystems, for bioaccumulative substances, and for cases in which an important species is not protected at the 5th percentile level. To provide further information to water quality managers in Australia/New Zealand, other percentile levels are also calculated so that criteria are given based on the 1st, 5th, 10th, and 20th percentiles.

Other researchers have also found good correlation between criteria derived from the 5th percentile of single-species SSDs and NOECs determined in multispecies tests (Maltby 2005; Hose and Van Den Brink 2004; Versteeg et al. 1999). In contrast, Zischke et al. (1985) found that a laboratory-derived criterion concentration of pentachlorophenol was not protective of invertebrates and fish in outdoor experimental channels. Using only arthropod species, Maltby et al. (2005) determined that concentrations of pesticides derived from the 5th percentile of SSDs (with 95% confidence) were protective of freshwater ecosystems whether pesticide applications were made once, multiple times, or continuously. Concentrations derived with 50% confidence were not protective for continuous or multiple pesticide applications; a safety factor was therefore required to arrive at no-effect concentrations.

The 5th percentile SSD cutoff, which has been validated against multispecies NOECs in several cases, is commonly used by current methodologies. It is a level that balances the desire to select a percentile near zero with the need to avoid the high uncertainty that exists in the extreme tails of the distributions.

Once a percentile is chosen, a decision on desired level of certainty for the resulting concentration must be made. The USEPA methodology (1985) does not provide a means to determine levels of confidence in the derived criteria. All other SSD methodologies result in a criterion derived from a specified percentile level and a specified level of confidence. Uncertainty in an extrapolated value results from a probability that the extrapolated value is wrong (Aldenberg and Slob 1993). The distribution around the extrapolated value can be used to calculate lower boundaries for the extrapolated value (Kooijman 1987; Van Straalen and Denneman 1989; Wagner and Løkke 1991; Aldenberg and Slob 1993). By evaluating this uncertainty, it is possible to state that the true HC_5 falls above (or below) the estimated value with a 50%, 90%, 95%, or other level of certainty. Among the calculated confidence levels, the most statistically robust is the 50th percentile, or median, estimate (ANZECC and ARMCANZ 2000; EVS Environmental Consultants 1999; Fox 1999). The variability in the tails of the distribution tends to compound, rather than clarify, the uncertainties.

The Dutch methodology (RIVM 2001) utilizes the 50% confidence, or median, HC_5 estimate for derivation of MPCs, but they also report a 90% two-sided confidence interval. Similarly, the Dutch methodology utilizes the median estimate of the HC_{50} for derivation of the SRC_{ECO}, but also report the 90% confidence interval. The Australia/New Zealand guidelines (ANZECC and ARMCANZ 2000) follow the Dutch example in using the median estimate of the HC_5 to derive the most probable estimate of the MTC. The Danish methodology, though, uses the lower 95th percentile estimate of the 5th percentile to derive criteria (although the Danish prefer to use an AF method; Samsoe-Petersen and Pedersen 1995). The EU risk assessment TGD utilizes the median PNEC estimate, but also considers the 95th percentile estimate in determining whether or not an AF should be applied to the derived PNEC. The OECD guidance (OECD 1995) allows calculation of either median or 95th percentile HC_5 estimates, and leaves it to the user to choose which level to use. Figure 1 depicts the median and lower 95th percentile estimates of the 5th percentile.

Maltby et al. (2005) investigated SSDs for pesticides and determined that the 95th percentile estimate of the SSD 5th percentile derived an HC_5 that was protective of ecosystems. The median 5th percentile level was protective for a single pesticide application, but was not protective of continuous or multiple applications. The authors suggest using a safety factor to address this shortfall. However, in this study, the SSDs were constructed from acute toxicity data, and it is not expected that an HC_5, derived from acute data, would be protective in continuous exposure scenarios. Multiple exposures would be better addressed by consideration of a frequency component in the criterion statement.

As discussed earlier, one challenge in the use of SSDs is to fit the data to an appropriate distribution before extrapolation. One way to achieve a better fit is to break data into groups rather than pooling them in one SSD. Data may be grouped according to toxicant mode of action, habitat (e.g., freshwater vs saltwater), reproductive strategy, or life cycle (Solomon and Takacs 2002). Newman et al. (2000) found that cumulative frequency models that did not fit log-normal or log-logistic models had distinct shifts in slope corresponding to transitions among taxa in the ranked data set. When data are grouped according to taxa or toxic mode of action, more data sets fit the log-normal distribution (ECOFRAM 1999; Newman et al. 2002). Traas et al. (2002) also support the idea of separating data into subgroups by taxa, or according to toxic mode of action, before constructing SSDs. In constructing SSDs for pesticides, Maltby et al. (2005) found that composition of taxonomic assemblages affected the hazard assessment, but groupings by habitat and geographic distribution had no effect.

Only the USEPA criteria methodology (1985) explicitly separates data into groups for constructing SSDs; moreover, the SSD is constructed using animal data only. Plants are indirectly included in criteria derivation. If a plant proves to be the most sensitive among tested species, then the FPV becomes the FCV. All other methodologies combine all aquatic data. The Netherlands methodology even includes NOECs derived from secondary poisoning analysis for birds and mammals (RIVM 2001). However, according to some of the guidelines, if statistical analysis shows that the data do not fit the assumed SSD distribution, or if data show a bimodal distribution, then data may be grouped to achieve a fit, with the most sensitive group used to derive the criterion (RIVM 2001; ECB 2003). In deriving target values using the Australia/New Zealand methodology (ANZECC and ARMCANZ 2000), which involves fitting data to one of several possible distributions, it was possible to use all data sets in their entirety (i.e., with all taxa combined).

Data have been grouped and/or excluded in other studies. For example, in constructing an SSD for an ecological risk assessment of chlorpyrifos, Giesy et al. (1999) excluded data from rotifers, mollusks, and other insensitive organisms, although such exclusions were not based on statistical analyses. Similarly, in a risk assessment of diazinon in the Sacramento and San Joaquin River basins, Novartis Crop Protection (1997) considered 10th percentile values for a combined fish and arthropod data set, as well as for separate fish and arthropod sets. The 10th percentile was derived from combined sets (3,710 ng/L), fish alone (79,900 ng/L) and for arthropods (483 ng/L). Using these numbers, the risk to arthropods would be underestimated if fish and arthropod data were combined, indicating that data for the two groups should be analyzed separately.

When the goal of a water quality criterion is to protect all species in an ecosystem, it is important to include all species in the derivation procedure. However, it is reasonable, especially in construction of SSDs, to separate species into groups if a multimodal distribution is evident. If there is no statistically significant difference between apparent groups (e.g., saltwater and freshwater, or plants and animals), then the data should be pooled for criteria derivation.

Details of the currently utilized SSD procedures are presented in the following paragraphs. Several countries use EPA methods and will not be individually treated. South Africa utilizes the USEPA methodology (1985). The OECD methodology (OECD 1995) refers users to the USEPA SSD procedures (1985), as well as those utilized (at the time) by the Netherlands (RIVM 2001) and Denmark (Samsoe-Petersen and Pedersen 1995).

USEPA (1985). To calculate the FAV, the GMAVs are ordered from highest to lowest and assigned ranks from 1 to N. For each GMAV, a cumulative probability (P) is calculated as $P = R/(N + 1)$. The four GMAVs nearest to $P = 0.05$ are selected (for data sets with fewer than 59 GMAVs, these will always be the four lowest values in the set). Using the selected GMAVs and P values, the FAV is calculated as follows (see Erickson and Stephan 1988 for derivation):

$$s^2 = \frac{\sum((\ln \mathrm{GMAV})^2) - ((\sum \ln \mathrm{GMAV}))^2/4}{\sum(P) - ((\sum(\sqrt{P}))^2/4)} \tag{2.7}$$

$$L = (\sum(\ln \mathrm{GMAV}) - s(\sum(\sqrt{P})))/4 \tag{2.8}$$

$$A = s(\sqrt{0.05}) + L \tag{2.9}$$

$$\mathrm{FAV} = e^A \tag{2.10}$$

where

s^2 = variance of lowest four values in the data set

GMAV= genus mean acute value

P= percentile rank of datum

L is as defined in Eq. (8)

A is as defined in Eq. (9)

FAV= final acute value

e= base of the natural logarithm

The acute criterion, called the CMC, is equal to the FAV/2.

The FCV may be derived in the same manner if enough chronic data are available; however, the FCV is typically derived by application of an ACR to the FAV. The chronic criterion, called the criterion continuous concentration (CCC), is the lowest value among the FCV, the FPV, or the FRV.

The Netherlands (RIVM 2001). ERLs are derived using the SSD method of Aldenberg and Jaworska (2000). That is, HC_p values are calculated based on a log-normal SSD. The HC_5 and HC_{50} are calculated as follows:

$$\log \mathrm{HC}_p = \overline{x} - k \cdot s \tag{2.11}$$

where

HC_p = Hazardous concentration for p% of species

$\overline{x}$ = mean of log-transformed NOEC data

k= extrapolation constant depending on percentile, level of certainty, and sample size (Table 1 in Aldenberg and Jaworska 2000)

s= standard deviation of log-transformed data

A computer program is available for making these calculations (RIVM 2004).

HC_5 values are used as MPCs, which are used to derive EQSs. A NC, which serves as an EQS target value, is equal to the MPC/2. HC_{50} values are used as SRC_{ECO}, which are EQS intervention values (i.e., the ecosystem is seriously threatened because 50% of species are adversely affected).

Denmark (Samsoe-Petersen and Pedersen 1995). The Danish methodology utilizes the SSD method of Wagner and Løkke (1991), which is essentially the same as that used in the Netherlands (RIVM 2001), but is stated differently and only calculates a lower one-sided confidence limit 5th percentile value. A value called a protection concentration (K_p) is calculated as follows:

$$K_p = \exp(\bar{x} - s \bullet k) \qquad (2.12)$$

where

K_p = concentration protecting (100-p)% of species with a specified level of confidence

p= percentile cutoff level

$\bar{x}$ = mean of log EC or log NOEC data

s= standard deviation

k = one-sided tolerance limit factor for a normal distribution depending on chosen confidence level (from Wagner and Løkke 1991)

The SSD method is only used in Denmark to estimate water quality criteria. An AF method is preferred, and is given more weight in deriving criteria.

Australia/New Zealand (ANZECC and ARMCANZ 2000). The Australia/New Zealand guidelines use the same method as do the Dutch, but with a curve-fitting procedure that overcomes the problem of data that do not fit an assumed distribution. Using the program BurrliOZ v. 1.0.13 (CSIRO 2001; Campbell et al. 2000), data are first fitted to one of a family of Burr distributions (Burr 1942; the log-logistic distribution is in the Burr family). After an appropriate distribution is chosen, then the calculation of the median HC_5 value is the same as shown for the Dutch methodology but utilizes extrapolation factors (k) derived for each of the distributions.

*EU Risk Assessment Guidelines (ECB*2003). Similar to the Australia/New Zealand approach, the EU TGD (ECB 2003) utilizes the Dutch SSD procedure, but with the provision that the distribution that best fits the data should be used. Either the Anderson-Darling or Kolmogorov-Smirnov test may be used to check goodness of fit. The PNEC is calculated as follows:

$$\text{PNEC} = \frac{5\%\ \text{SSD}(50\%\ \text{c.i.})}{\text{AF}} \qquad (2.13)$$

where

PNEC= predicted no-effect concentration

5% SSD = concentration determined from SSD expected to protect 5% of species

50% c.i.= 50% confidence interval

AF= assessment factor of 1–5

When enough data are available, SSD methodologies provide a reasonable way to estimate ecosystem-level effects based on single-species data. Several criticisms have been directed at SSDs and their use in setting regulatory limits. Most of the criticism stems from the underlying assumptions in SSD methodologies, some of which are discussed in the OECD (1995) and Australia/New Zealand guidelines (ANZECC and ARMCANZ 2000). The most general of these assumptions are discussed here. First, is the assumption that the ecosystem is protected if 95% of species in the ecosystem are protected. This assumption may be particularly problematic if so-called keystone species are among the most sensitive to a toxicant. Any criterion derived by any method must be evaluated in the context of species considered to be important for ecological, commercial or recreational reasons. If data indicate that important species will be harmed by the derived criterion, then an adjustment of the criterion is in order. The USEPA methodology (1985) stipulates that, if a species mean acute (chronic) value (SMAC or SMCV, respectively) of a commercially or recreationally important species is lower than the calculated FAV (FCV), then the SMAC (SMCV) is used as the FAV (FCV).

Another assumption discussed by the OECD (1995) and the Australia/New Zealand (ANZECC and ARMCANZ 2000) guidances is that the distribution of toxicity data is symmetrical. If insensitive species give very high toxicity values and sensitive species give very low values, then this bimodal distribution will probably have a very large standard deviation. The large standard deviation, resulting from such data, will portend a very low estimate of the 5th percentile level. The best way to handle bimodal distributions is to scrutinize data and determine if outliers should be removed from the set (ANZECC and ARMCANZ 2000); alternatively, data can be split into two distributions and the more sensitive data used to derive criteria (RIVM 2001; ECB 2003). A third assumption (ANZECC and ARMCANZ 2000; OECD 1995) is that toxicity data represent independent, random samples from the distribution, which is generally not true; data are normally collected from species that are easy to handle in the laboratory, or were selected for their sensitivity to a particular toxicant. For some species, there are many data and for some there are none.

Posthuma et al. (2002a, b) point out a number of advantages, disadvantages, and ongoing issues in the use of SSD methods. Advantages include (1) SSD methods are conceptually more transparent and scientifically more defensible than AF methods; (2) they are widely accepted by regulators and risk assessors; (3) they are understandable; (4) they allow risk managers to choose appropriate percentile levels and confidence levels; (5) they use commonly available ecotoxicity data; (6) they rely on relatively simple statistical methods; (7) they provide a way to assess mixtures; (8) they can be used to determine effects on species or on communities; and (9) they provide clear graphical summaries of assessment results. Disadvantages include (1) SSD methods have not been proven to be more (or less) reliable than alternatives; (2) they require relatively large data sets; (3) they rely on statistics with no mechanistic components; (4) distributional assumptions may not be true; (5) multimodal species distributions are problematic; (6) criteria based on lower confidence limits are overprotective; (7) test species are not randomly sampled; (8) there is no weighting of important species; (9) sensitive species may be overrepresented; (10) important species may fall in the unprotected range; and (11) ecosystem functions are not represented. Ecological issues discussed by Posthuma et al. (2002a) include the problem of using data from a few species in laboratory conditions to represent responses of many species under field conditions. The authors note that laboratory data are often biased toward very sensitive or very tolerant species, and are from studies conducted under conditions that do not account for bioavailability and multiple routes of exposure. Statistical issues include choice of toxicological endpoint, data set distribution type, choice of percentile level to represent a no effect, and methods of quantifying uncertainty.

Despite violations of some of the assumptions, and despite the disadvantages, SSD methods have many advantages over AF methods in criteria derivation. Particularly, important is the ability for risk managers to select appropriate percentile levels and confidence levels, which is not possible by the AF method. So far, criteria derived from SSDs have proven to be protective of ecosystems. Further future validation will take place as the database of field studies expands (OECD 1995).

7.3 Other Considerations in Criteria Derivation

7.3.1 Mixtures

A recurring criticism of deriving water quality criteria from singles-species, single-chemical laboratory toxicity tests is that such tests do not account for the multiple stressors facing organisms in the field. In the environment, organisms must deal with chemical mixtures, physical stressors, and interactions with other organisms. Methods to incorporate the effects of temperature, pH, and other environmental factors into criteria derivation have been discussed. Species interactions can only be addressed in multispecies toxicity tests. This section specifically addresses the effects of contaminant mixtures.

Results of stream monitoring in the US revealed that more than 50% of samples contained five or more pesticides (USGS 1998). The California DPR reports that over 175 million pounds of hundreds of different pesticides were commercially applied in California in 2003 (California DPR 2005b). It is, therefore, probable that various mixtures will be present in surface waters as a result of transport processes such as drift and runoff. Studies of the effects of mixtures are few and represent an extremely small portion of the number of mixtures that could potentially occur in the environment. Water quality criteria, derived from single-chemical exposures, have proven to be protective of ecosystems, but a key question is, if chemical A and B show additive (or synergistic or antagonistic) toxicity, then what level of each is environmentally acceptable. Lydy et al. (2004) discuss the challenges of regulating pesticide mixtures, considering our limited knowledge of pesticide interactions. Alabaster and Lloyd (1982) report that joint toxicity of pesticide mixtures is more than additive in a high proportion of cases; moreover, they demonstrate this characteristic more often than do other kinds of toxicants. In contrast, Mount et al. (2003) believe that very few cases of extreme antagonism or synergism are observed in environmental mixtures and, therefore, an assumption of only additivity is appropriate in most cases of chemical mixtures. Warne and Hawker (1995) proposed the funnel hypothesis, which states that deviations from additivity in mixtures, decrease with increasing numbers of components in the mixture, thus for very complex mixtures, additivity models are likely to be valid.

There are two models used to assess additive toxicity. One is the response addition model, in which the chemicals have different modes of action and do not interact with each other. The second is the concentration addition model, in which chemicals have the same mode of action, but do not interact with each other (Plackett and Hewlett 1952). According to Mount (2003), the response addition model is not widely accepted, because it cannot be readily tested. However, the concentration addition model has been successfully tested for several modes of action and may be used to derive technically defensible criteria.

The concentration addition model is applied in the Water Quality Control Plan (Basin Plan) for the Sacramento River and San Joaquin River basins (CVRWQCB 2004). In the Basin Plan, if multiple chemicals with similar modes of action are present in a water body, WQOs are met if the following is true:

$$\sum_{i=1}^{n} \frac{C_i}{O_i} \leq 1.0 \qquad (2.14)$$

where

C_i = concentration of toxicant i in water

O_i = WQO for toxicant i.

In reviewing proposed Basin Plan amendments, Felsot (2005) noted that, for diazinon and chlorpyrifos, in particular, this additivity analysis is not appropriate because the denominator is based, not on actual toxicity values, but on an objective that includes a safety factor of 2. He proposes that a better way to determine added toxicity compliance is to use the relative potency factor (RPF) approach, which is analogous to the toxic equivalency factor (TEF) approach used in assessing the toxicity of dioxin and dioxin-like compounds. Using the RPF approach, one chemical (usually the most toxic one) is chosen to be the reference chemical, and the potency of other similarly acting chemicals is expressed as a ratio to the reference chemical's toxicity. This ratio, the RPF, is multiplied by measured concentrations of each nonreference chemical to produce concentrations expressed as equivalents to the reference chemical. Compliance with the objective for the reference chemical is based on the sum of the measured reference chemical(s) plus the concentrations of those expressed in its equivalents.

The USEPA guidelines (1985) do not incorporate mixtures or multiple stressors into aquatic life criteria derivation. However, the regulators of the Central Valley RWQCB may use mixture models to assess compliance with objectives. Similarly, the Australia/New Zealand guidelines do not derive mixture criteria, but determine compliance using the following formula (ANZECC and ARMCANZ 2000):

$$\mathrm{TTM} = \Sigma(C_i / \mathrm{WQG}_i) \tag{2.15}$$

where

TTM = total toxicity of the mixture

C_i = concentration of the *i*th component of the mixture

WQG_i = water quality guideline for that component

If the TTM exceeds 1.0, then the water quality guideline has been exceeded.

For more complex mixtures (>5 components), the Australia/New Zealand guidelines prefer the technique of DTA of effluents and receiving waters (equivalent to whole effluent or ambient toxicity testing in the US). DTA is a good tool to determine if waters are able to support aquatic communities, but without follow-up toxicity identification evaluation, it does not provide information to define what chemical or chemicals, in a mixture, may cause any observed toxicity. Again, this is a monitoring and compliance tool, rather than a way to address derivation of criteria for mixtures.

The Dutch, Danish, OECD, South African, Canadian, EU, Spanish, and French methodologies do not directly address mixtures or multiple stressors (RIVM 2001; Samsoe-Petersen and Pedersen 1995; Roux et al. 1996; CCME 1999; Bro-Rasmussen et al. 1994; Lepper 2002). The EU risk assessment TGD, and the German guidelines include mixture effects on lists of uncertainties that may be addressed by using AFs, but offers no further guidance on mixtures (ECB 2003; Irmer et al. 1995). In the UK, combined EQSs may be derived for structurally similar substances with similar modes of action (Zabel and Cole 1999). Lepper (2002) proposes a method for derivation of quality standards that does not explicitly account for the toxicity of mixtures, but does utilize the AF method, described in the EU risk assessment TGD (ECB 2003), which includes factors for mixture effects.

Another way to address mixtures is provided by Könemann (1981). He developed a maximum toxicity index (MTI), which can be used to quantify toxicity of mixtures of two or more chemicals that have either similar or independent action. Könemann's model and the others discussed above, only work when trying to determine if toxicity is additive, more than additive, or less than additive. Rider and LeBlanc (2005) have recently proposed a model that incorporates toxicokinetic chemical interactions, as well as concentration and response addition. This model, called the integrated addition and interaction (IAI) model, correctly predicted joint toxicity of 30 ternary mixtures containing known interactive chemicals. Models that assumed no interaction did not accurately predict joint toxicity.

Finally, SSDs offer a means of assessing mixture toxicity. Traas et al. (2002) discussed how to determine a multisubstance potentially affected fraction (msPAF; essentially the same as the HC_p) for cases of concentration addition (i.e., similar mode of action), and response addition (i.e., different modes of action). The calculations are somewhat complicated, and the reader is referred to Traas et al. (2002) for details. To calculate an overall msPAF, SSDs are first calculated individually for each chemical in the mixture. The concentration addition calculation method is then applied to groups with similar modes of action, yielding an msPAF for each mode of action in the mixture. The individual PAF values that did not fit into any groups, based on mode of action, are aggregated with the msPAFs by the response addition calculation method, to yield an overall msPAF for the mixture. It is current practice to apply the concentration addition calculation method to groups of chemicals with narcotic modes of action, and to photosynthesis inhibitors and acetylcholinesterase inhibitors (Traas et al. 2002). Typically, this aggregation would be performed within taxonomic groups, but Posthuma et al. (2002b) extend this to suggest that it is possible to aggregate across taxa to derive an msPAF for all species in an ecosystem. By this method, an msPAF value less than 0.05 would indicate compliance with water quality criteria, whereas, an msPAF > 0.05 would indicate noncompliance (assuming criteria were derived to protect 95% of species).

Using the above methods, mixture toxicity is addressed at the compliance stage, not at the criteria derivation stage. Although it would be ideal to actually derive criteria for mixtures, it would be impossible to develop criteria for all potential pesticide mixtures that could occur in a water body. To determine compliance of criteria for individual chemicals, an appropriate model should be selected. If little is known about the actions and interactions of the chemicals in a mixture, then an additive assumption is reasonable and simple models may be used. However, if interactions are known to occur that lead to antagonistic or synergistic action, then a more complex model, such as that of Rider and LeBlanc (2005), should be used.

7.3.2 Bioaccumulation

Bioaccumulative chemicals pose risks that are not measured in standard laboratory toxicity tests. For bioaccumulative chemicals, many methods exist to incorporate bioaccumulation data into criteria derivation. Such methods may be simple, such as adjusting the size of the applied AF (Zabel and Cole 1999; Samsoe-Petersen and Pedersen 1995; Bro-Rasmussen et al. 1994; Lepper 2002), or using a tissue residue level to determine a chronic criterion (USEPA 1985). They may also involve converting food-based NOECs for fish-eating predators into water-based NOECs, which can be combined with other water-effects data, in criteria derivation (RIVM 2001; OECD 1995). Others methods do not address bioaccumulation in aquatic life water quality criteria, but do so in other ecological effect assessments (CCME 1999; USEPA 2003a). For example, the Great Lakes guidance includes a procedure for derivation of water quality criteria for the protection of wildlife (USEPA 2003a). The South African methodology does not consider bioaccumulation at all (Roux et al. 1996).

The FRV used in the USEPA methodology (1985) is intended to prevent exceedance of FDA action levels in recreationally or commercially important species, and to protect wildlife, including fishes and other animals that consume aquatic organisms, where adverse dietary effects have been demonstrated. The FRV is a water concentration derived by dividing a maximum permissible tissue concentration by a BCF (uptake directly from water) or BAF (uptake from water and food). BAFs are preferred for the FRV calculation, but because BAFs are generally not available, BCFs are often used. The maximum permissible tissue concentration may constitute an FDA action level for fish oil, or the edible portion of fish or shellfish, or a maximum dietary intake level that will not cause adverse effects on survival, growth, or reproduction. If multiple BCF values are available, the highest geometric mean species BCF is used. For protection of fish-eating wildlife, the BCF should be based on whole-body measurements, while for human health concerns it should be based on the edible portion of the fish (whole fish in some cases and for some cultures). The FRV is selected as the lowest of all residue values determined (for different species, including humans). If the FRV is the lowest of the FCV, the FRV and the FPV, the chronic criterion is set at the FRV. Similarly, for protection of fisheries (human consumption), the German approach is to develop water quality targets by dividing the allowable food residue by a BCF (Irmer et al. 1995; BMU 2001).

Chemicals that do not pose a risk to primary producers or consumers may pose risks to organisms, particularly terrestrial organisms, higher up the food chain, if those chemicals have the potential to bioaccumulate. In the Netherlands methodology, this is addressed via consideration of secondary poisoning (RIVM 2001). For substances with log $K_{ow} > 3$, molecular weight <700, low metabolism or excretion rate and/or other literature evidence of bioaccumulative potential, secondary poisoning must be considered in deriving criteria. BCFs or BAFs (if available) are used to convert predator NOECs to water NOECs as follows:

$$\text{NOEC}_{\text{water,fish-to-predator}} = \frac{\text{NOEC}_{\text{predator}}}{\text{BCF}_{\text{fish}}} \times 0.32 \tag{2.16}$$

$$\text{NOEC}_{\text{water,mussel-to-predator}} = \frac{\text{NOEC}_{\text{predator}}}{\text{BCF}_{\text{mussel}}} \times 0.20 \tag{2.17}$$

where 0.32 and 0.20 are factors to correct for caloric content of food.

These converted NOECs are combined with all other aquatic effects data from DTAs to calculate an ecosystem MPC (MPC_{ECO}). MPCs are also calculated separately for predators, and for the aquatic compartment and the independently derived values are reported for comparison.

The OECD (1995) also provides guidance for consideration of secondary poisoning. According to this methodology, chemicals are likely to bioaccumulate if they have $K_{ow} > 3$, molecular weight <1000, molecular diameter <5.5 Å, and molecular length <5.5 nm. Reactive and readily metabolized substances are not expected to bioaccumulate. The OECD requires that BCFs be expressed on a whole body fresh, or wet weight basis and that they not be lipid normalized. BCFs may be either measured experimentally or may be estimated using the K_{OW}. The OECD determined that secondary poisoning risks to predatory fish are not of concern, after reviewing several modeling studies (OECD 1995 based on Barber et al. 1988; Gobas et al. 1988; Norstrom et al. 1976; Thomann and Connolly 1984). Therefore, only secondary poisoning in fish-eating mammals and birds is addressed. Unfortunately, the OECD guideline authors seem to have misinterpreted these studies. For example, Thomann and Connolly (1984) found that dietary uptake of PCBs accounted for 99% of body burden in adult trout. Moreover, Gobas et al. (1988)

compiled data showing that average efficiency of absorption of hydrophobic organic chemicals from food for salmon and rainbow trout was 0.45 ± 0.06, for chemicals with log K_{ow}s < 7.0, and was 0.18 ± 0.04 for chemicals with log K_{ow}s ≥ 7.0. Thus, dietary uptake may be less important for extremely hydrophobic chemicals, but it is prospectively important as an exposure route. The OECD guidelines use toxicity data to derive a maximum concentration in food that will minimize risk for fish-eating wildlife (other than fish). This maximum concentration is divided by the BCF, for fish, to give the MTC for water (based on a method presented by Romijn et al. 1993). If the water MTC, derived in this manner, is lower than the MTC derived for protection of aquatic life, then secondary poisoning must be considered in setting criteria.

The EU risk assessment TGD (ECB 2003) describes potentially bioaccumulative chemicals as those that have a log K_{ow} > 3, *or* are highly adsorptive, *or* belong to a class of chemicals known to be bioaccumulative, *or* have a structure that indicates bioaccumulative potential, *and* have no features that might mitigate bioaccumulative potential (e.g., short half-life). The TGD provides guidance for assessment of secondary poisoning, but from the perspective of assessing risk to predators from dietary uptake, rather than for setting water quality criteria. That is, BCFs and BMFs (relative concentration in predator compared to prey) are used to predict concentrations of contaminants in prey, based on concentrations measured in water, using the following equation:

$$\mathrm{PEC}_{\mathrm{oral,predator}} = \mathrm{PEC}_{\mathrm{water}} \times \mathrm{BCF}_{\mathrm{fish}} \times \mathrm{BMF} \qquad (2.18)$$

where

$\mathrm{PEC}_{\mathrm{oral,predator}}$ = predicted environmental concentration in food

$\mathrm{PEC}_{\mathrm{water}}$ = predicted environmental concentration in water (from exposure assessment)

$\mathrm{BCF}_{\mathrm{fish}}$ = BCF for fish on wet weight basis

BMF= biomagnification factor in fish

By substituting NOEC values for the PEC values in Eq. (18), one could solve for $\mathrm{NOEC}_{\mathrm{water}}$ given a $\mathrm{NOEC}_{\mathrm{oral,predator}}$, BCF and BMF values:

$$\mathrm{NOEC}_{\mathrm{water}} = \frac{\mathrm{NOEC}_{\mathrm{oral\text{-}predator}}}{\mathrm{BCF}_{\mathrm{fish}}\mathrm{BMF}} \qquad (2.19)$$

Inclusion of the BMF value accounts for residues in fish, which result both from direct water uptake and dietary intake of contaminated organisms from lower trophic levels. If experimentally determined BMF values are not available, default values taken from other studies are used in the TGD. As noted, the USEPA (1985), the OECD (1995) and the RIVM (2001) guidelines discuss BAFs (which include dietary uptake), but use BCFs (no dietary component) in their calculations, because BCFs are normally available. However, these methodologies make no attempt to correct the calculated water concentrations for the dietary uptake component. This results in water concentration values (e.g., FRVs in the case of USEPA) that are higher than needed, because the BMF term in the denominator is lacking and, if present, would reduce the $NOEC_{water}$.

The Canadian water quality criteria derivation protocol (CCME 1999) does not add an additional factor for bioaccumulation. However, to derive a full guideline for a bioaccumulative chemical does require reporting of bioaccumulation data; if such data are lacking, then only an interim guideline can be derived. Moreover, Canada has separate tissue residue guidelines (TRGs) for protection of fish-eating wildlife (CCME 1997). These TRGs are not translated into safe water levels, but are expressed as safe levels in fish tissue. Many TRGs are based on human health studies, because data on safe dietary levels of chemicals in wildlife are unavailable.

Similar to the Canadian approach, the USEPA Great Lakes guidance (USEPA 2003a) does not incorporate bioaccumulation into aquatic life criteria. Rather, it provides for derivation of separate water quality criteria for the protection of wildlife and human health. The end result is that separate water quality criteria exist for aquatic life, wildlife, and human health. Presumably, the decision about which one(s) to apply to a particular water body, depends on beneficial use designations.

The authors of the Australia/New Zealand guidelines (ANZECC and ARMCANZ 2000) chose not to incorporate bioaccumulation into water quality criteria derivation guidelines for two major reasons. First, the authors determined that the link between concentrations of bioaccumulative chemicals in water and secondary poisoning is not strong. Second, they determined that there is insufficient international guidance for deriving bioaccumulation-based criteria. To address uncertainty surrounding what constitutes a safe level for bioaccumulative chemicals (chemicals with log K_{ow} values between 3 and 7), when setting criteria for them, the Australia/New Zealand guidelines use the 1st percentile of the species distribution (rather than the 5th). Also, the guidelines allow for site-specific, case-by-case application of available methods for translating wildlife dietary levels into water quality criteria.

Because there is a potential for secondary poisoning effects in aquatic and terrestrial animals, as well as human health concerns (which affect commercially and recreationally important species), it is important to include a method of incorporating bioaccumulation into water quality criteria derivation. If a linkage can be made between dietary exposure and adverse effects (e.g., from wildlife studies or FDA action levels) then those effects data should be used, along with BCFs, BAFs, and BMFs to translate food item concentration limits into water concentrations.

7.3.3 Threatened and Endangered Species

Because of their protected status, it is probable that very little toxicity test data will exist for TES. However, it is important to ensure that these species are protected by water quality criteria. Setting national criteria for TES that have limited geographic range makes little sense, which explains why very few of the national criteria derivation methodologies address TES. However, if the goal of a project is to develop regional criteria, then protection of TES should be considered.

Among the few methodologies that do address TES, procedures are given for adjustment of criteria on a site-specific basis, provided data are available to assure that one or more TES may not be protected by the criterion. For example, the Great Lakes guidance (USEPA 2003a) provides two methods for altering criteria that may protect TES (listed or proposed for listing): (1) if the SMAV for the TES, or a surrogate species is lower than the FAV, then that SMAV may be used as the FAV; or (2) site-specific criteria may be derived using the criteria recalculation procedure, described in the USEPA *Water Quality Standards Handbook* (USEPA 1994). The Australia/New Zealand guidelines suggest that TES may be protected through selection of surrogate species (appropriate to a particular site) for inclusion in the data set used to derive TVs (ANZECC and ARMCANZ 2000).

The ICE and QSAR approaches provide for quantitatively estimating toxicity for TES, based on toxicity to surrogate species. While QSARs are limited to a few specifically acting substances, ICE models can be applied to any substance. In contrast, the ICE model has only been developed for acute toxicity, whereas QSARs exist for prediction of both acute and chronic toxicity. These two estimation techniques would probably best be used as a means to assess potential harm to TES by comparing estimated toxicity values to derived criteria.

7.3.4 Harmonization/Coherence Across Media

The concept of harmonization is aptly described as follows (RIVM 2001): "The objective of the harmonization procedure is to compare the concentrations at steady state in the receiving compartments … with the MPCs that have been derived for these compartments from the (eco)toxicological data. If this comparison indicates that maintaining the concentration in the primary compartment (the compartment of emission) at MPC level results in exceeding the MPC in any of the secondary compartments, the set of MPCs must be considered incoherent and has to be adjusted." Briefly, according to the Dutch methodology (RIVM 2001) the scheme for harmonizing aquatic life water/sediment/soil ERLs is as follows: (1) if sufficient direct toxicity data are unavailable for soil or sediment, then the ERL is derived from water data by the equilibrium partitioning (EqP) method, and this becomes the final, harmonized soil/sediment ERL; (2) if there is sufficient data to allow statistical extrapolation (by a refined effects assessment) of soil/sediment data, then the ERL is derived directly and no further harmonization is required; (3) if the soil/sediment ERL is determined by a preliminary effects assessment (by application of AFs to limited data sets), then this value is compared to the value determined by the EqP approach and the lower of the two values is accepted as the harmonized ERL. This procedure is used with the caveat that there are uncertainties in both the ERLs and in the partition coefficients.

To harmonize ERLs with human health risk limits, the Netherlands methodology (RIVM 2001) uses a multimedia box model to estimate equilibrium concentrations in secondary (receiving) compartments, providing an ERL has been derived for a primary (emission) compartment. Utilizing Van De Meent's (1993) SimpleBox environmental fate model, it is possible to determine if an ERL, derived for a primary compartment, has the potential to result in exceedance of an ERL, or human health limit, in another compartment. It is not clearly explained why this approach is not used to harmonize water, soil, and sediment ERLs, though it appears that this type of model could be used to harmonize aquatic life criteria across environmental compartments.

In the German methodology, the most sensitive asset (e.g., drinking water, aquatic life) is taken as the basis for deriving the WQO (BMU 2001). For example, if the drinking water target for a substance is 0.1 μg/L, and the aquatic life target is 0.05 μ/L, then the aquatic life target becomes the objective.

Cross-media coherence of criteria is addressed by only the few methodologies mentioned here. Lack of attention to this issue is probably because of gaps in knowledge and paucity of data for development of models to describe cross-media processes. Benson et al. (2003) noted that models have been successfully used for assessing possible conflicts between water and sediment criteria for some compounds, but fully integrated quantitative multimedia models are not available for making full intermedia assessments. Although it may not be possible to derive fully integrated criteria, it is important to use available models to determine if excursions above water quality criteria might adversely affect other environmental compartments.

7.3.5 Utilization of Available Data and Encouragement of Data Generation

Many methodologies make very poor use of available data, because they use only the lowest values (Lepper 2002; CCME 1999) or focus on the lowest few values in a data set (USEPA 1985; Roux et al. 1996). The SSD methodologies, utilized in the Netherlands (RIVM 2001) and Australia/New Zealand (ANZECC and ARMCANZ 2000), make full use of data. Among their good use of data is the fact that they utilize variability information to derive confidence limits for criteria. In particular, the Australia/New Zealand curve-fitting method reduces the need to remove outliers or truncate data sets that show a degree of multimodality.

A recurring theme in this chapter is that ecotoxicity data are generally too scarce. Often, insufficient ecotoxicology data are available to derive adequate criteria (where "adequate" means there will be high certainty that criteria will neither over- nor under-protect aquatic ecosystems). Therefore, it would be beneficial if a criteria derivation methodology was designed to encourage data generation by all stakeholders. Okkerman et al. (1991) found an example in which HC_5 values, based on data for five species, were lower than those based on nine species. Such examples exist because the uncertainty in the SSD method decreases with increasing sample size (which results in lower standard deviations and extrapolation factors).

In contrast, for the USEPA method (1985) which uses only the four values nearest the 5th percentile (the lowest four values in many cases) to calculate the FAV, additional data may have different effects on the FAV, depending upon whether the new data fall within the group of four nearest the 5th percentile. This is illustrated in a report prepared for the California State Water Resources Control Board by the Great Lakes

Environmental Center (GLEC 2003). In Appendix C of that report, the authors present results of various manipulations of a basic data set. First, with no change to the four values (data points) used to calculate the FAV, simply increasing the number of samples (*N*), always increases the FAV, as a consequence of the variability in *P* values of the four data points being reduced. Second, as the range of the four values increases (i.e., the variability among the four data points increases) the FAV decreases, because of the increased variability around the 5th percentile. The problem with the first of these kinds of data set manipulations is that, in an effort to derive higher criteria by the USEPA method, one could simply conduct more tests with insensitive species. Aside from causing the set to violate the log-triangular distribution assumption, such data would drive the criterion upward in a predictable manner, based solely on *N*, because the new data would not be near the 5th percentile. With other SSD methodologies (i.e., those that do not ignore the upper part of the distribution), the best way to drive a criterion higher is to have a large, balanced data set, such that the variability in the whole set is reduced. By these other methods, if a data set were "padded" with extremely high or low values, outliers and bimodal distributions would be detected, and the set would be modified to fix these problems before the SSD analysis (ANZECC and ARMCANZ 2000; RIVM 2001; ECD 2003). To encourage generation of balanced data sets, SSD methods that utilize all data (RIVM 2001; ANZECC and ARMCANZ 2000) are preferable to those that focus on a limited range of the distribution.

Manufacturers and other waste dischargers have little incentive to generate data if an AF method is used to derive criteria, because new data that show no or low effects are ignored, while those that show high sensitivity drive a standard lower (Whitehouse et al. 2004). This is because only data from the most sensitive species are used to set criteria by AF methods. However, to the extent that AF methods utilize variable factors, they do foster data generation, in that factors are smallest for the most complete data sets, and smaller factors yield higher criteria values.

According to the Australia/New Zealand methodology, a high reliability TV can be established on the results of three high quality field or mesocosm studies. There is no stipulation that such TVs will only be used if they are lower than those derived by extrapolation methods; multispecies research is therefore encouraged (ANZECC and ARMCANZ 2000). In contrast, if adequate field or mesocosm data are available that indicate a FCV should be lower than the one calculated by the USEPA methodology (1985), then the FCV can be adjusted. However, this does not encourage generation of field or semi-field data by all stakeholders, because the FCV can only be adjusted downward, in this scenario.

8 Summary

Environmental regulators charged with protecting water quality must have scientifically defensible water quality goals. For protection of aquatic life, regulators need to know what levels of contaminants a water body can tolerate, without producing adverse effects. The USEPA has developed water quality criteria for many chemicals, but few are for current-use pesticides. Other countries also derive aquatic life criteria utilizing a variety of methodologies. As a prelude to developing a new criteria derivation methodology, this chapter explores the current state of aquatic life criteria derivation around the world. Rather than discussing each methodology independently, this review is organized according to critical elements that must be part of a scientifically defensible methodology.

All of the reviewed methodologies rely on effects data to derive aquatic life criteria. Water quality criteria may be derived from single-species toxicity data by statistical extrapolation procedures (for adequate data sets), or by use of empirically based AFs (for data sets of any size). Assessment factor methods are conservative and have a low probability of underestimating risk, with a concomitant high probability of overestimating risk. Extrapolation methods may also under-, or overestimate risk, but uncertainty is quantifiable and is reduced when larger data sets are used. Although less common, methods are also available for criteria derivation using multispecies toxicity data.

Environmental toxicity of chemicals is affected by several factors. Some of these factors can be addressed in criteria derivation, and some cannot. For example, factors such as magnitude, duration and frequency of exposure may be incorporated into criteria, either through use of time-to-event and population models or by derivation of both acute and chronic criteria that have duration and frequency components. Aquatic species may be exposed to hydrophobic organic chemicals by multiple routes. They may take up residues directly from water, or may be exposed dietarily, or combinations of both. Unfortunately, to properly address such multiple routes in criteria derivation, food web models are needed that work for chemicals that have specific modes of action. Similarly, both bioavailability and toxicity parameters may contribute to derivation of criteria, providing sufficient data are available.

Ecotoxicological effects and physical–chemical data are needed for criteria derivation. The quality and quantity of required data are clearly stated in existing methodologies; some guidelines provide very specific data quality requirements. The level of detail provided by guidelines varies among methodologies. Most helpful are those that provide lists of acceptable data sources, descriptions of adequate data searches, schemes for rating ecotoxicity data, specifications of required data types (e.g., acute vs chronic), and instructions for data reduction. Many methodologies present procedures for deriving criteria from both large and small data sets. Very small data sets may be supplemented through the use of QSARs for selected pesticides, and through the use of models such as ICE (for prediction of toxicity to under-tested species), and ACE (for estimation of chronic toxicity from acute data).

The toxicity of mixtures is addressed by several existing methodologies. In some cases, additional AFs are applied to criteria to account for exposure to mixtures, whereas in others, concentration addition models are used to assess compliance. Multiple stressors and bioaccumulation are also addressed in some methodologies, by providing for application of additional safety factors. Methods are also available for translating dietary exposure limits for humans or other fish-eating animals into water concentrations. Options for addressing the safety of TES exist, and rely heavily on data from surrogate species to derive criteria. Utilizing partition coefficients, criteria may be harmonized across media to ensure that levels set to protect one compartment do not result in unacceptable levels in other compartments.

Several methodologies derive criteria from entire data sets through the use of statistical extrapolations; other methods utilize only the lowest (most sensitive) data point or points. Utilization of entire data sets allows derivation of confidence limits for criteria, and encourages data generation.

Criteria derivation methodologies have improved over the past two decades as they have incorporated more ecological risk assessment techniques. No single existing methodology is ideal, but elements of several may be combined, and when used with newer risk assessment tools, will produce more usable and flexible criteria derivation procedures that are protective.

Acknowledgments We thank the following reviewers: Lawrence R. Curtis (Oregon State University), Brian Finlayson (California Department of Fish and Game), Evan P. Gallagher (University of Washington), John P. Knezovich (Lawrence Livermore National Laboratory), and Marshall Lee (California Department of Pesticide Regulation). This project was funded through a contract with the Central Valley RWQCB. Mention of specific products, policies, or procedures does not represent endorsement by the Regional Board. The contents also do not necessarily reflect the views or policies of the USEPA nor does mention of trade names or commercial products constitute endorsement or recommendation for use.

References

Alabaster JS, Lloyd R (1982) Water quality criteria for freshwater fish. Butterworth Scientific, Surrey, UK, pp 253–314.

Aldenberg T, Jaworska JS (2000) Uncertainty of the hazardous concentration and fraction affected for normal species sensitivity distributions. Ecotox Environ Saf 46: 1–18.

Aldenberg T, Slob W (1993) Confidence limits for hazardous concentrations based on logistically distributed NOEC toxicity data. Ecotox Environ Saf 25: 48–63.

Aldenberg T, Luttik R (2002) Extrapolation factors for tiny toxicity data sets from species sensitivity distributions with known standard deviation. In: Species Sensitivity Distributions in Ecotoxicology. Posthuma L, Suter IIGW, Traas TP (eds), Lewis Publishers, CRC Press, New York, NY.

ANZECC and ARMCANZ (2000) Australian and New Zealand guidelines for fresh and marine water quality. Australian and New Zealand Environment and Conservation Council and Agriculture and Resource management Council of Australia and New Zealand, Canberra, Australia.

Applegate JS (2000) The precautionary preference: An American perspective on the precautionary principle. Human Ecol Risk Assess 6: 413–443.

AQUIRE (1981-present) AQUIRE database. US Environmental Protection Agency. Available through ECOTOX at http://www.epa.gov/ecotox/.

AQUIRE (Aquatic Toxicity Information Retrieval Database) (1994) AQUIRE standard operating procedures. USEPA, Washington, DC.

Auer CM, Nabholz JV, Baetcke KP (1990) Mode of action and the assessment of chemical hazards in the presence of limited data: Use of Structure-Activity Relationships (SAR) under TSCA, Section 5. Environ Health Perspect 87: 183–197.

Bailey HC, Deanovic L, Reyes E, Kimball T, Larson K, Cortright K, Connor V, Hinton DE (2000) Diazinon and chlorpyrifos in urban waterways in Northern California, USA. Environ Toxicol Chem 19: 82–87.

Barber MC, Suarez LA, Lassiter RR (1988) Modeling bioconcentration of nonpolar organic pollutants by fish. Environ Toxicol Chem 7: 545–558.

Bedaux JJM, Kooijman SALM (1993) Statistical analysis of bioassays, based on hazard modeling. Environ Ecol Stat 1: 303–314.

Benson WH, Allen HE, Connolly JP, Delos CG, Hall LW Jr, Luoma SN, Maschwitz D, Meyer JS, Nichols JW, Stubblefield WA (2003) Exposure Analysis. In: Reevaluation of the State of the Science for Water-Quality Criteria Development, Reiley M, Stubblefield WA, Adams WJ, Di Toro DM, Hodson PV, Erickson RJ, Keating FJ Jr (eds), SETAC Press, Pensacola, FL.

BIODEG (1992) Biodegradation probability program (version 3.0). Now called BioWin. Available at http://www.syrres.com/esc/est_soft.htm.

BMU (2001) Environment Policy, Environment Resources Management in Germany, Part II, Quality of Inland Surface Waters. Federal Ministry for the Environment, Nature Conservation and Nuclear Safety, Div. WAI 1(B), Postfach 12 06 29, Bonn Germany.

Bockting GJM, Van De Plassche EJ, Struijs J, Canton JG (1993) Soil-water partition coefficients for organic compounds. RIVM Report No. 679101013. National Institute of Public Health and the Environment, Bilthoven, The Netherlands.

Borthwick PW, Clark JR, Montgomery RM, Patrick JM Jr, Lores EM (1985) Field confirmation of a laboratory-derived hazard assessment of the acute toxicity of fenthion to pink shrimp, *Penaeus duorarum*. In: Aquatic Toxicology and Hazard Assessment: Eighth Symposium. ASTM STP 891, Bahner RC, Hansen DJ (eds), American Society of Testing and Materials, Philadelphia, PA, pp 177–189.

Bringmann G, Kühn R (1977) Befunde der Schadwirking wassergefährdender Stoffe gegen *Daphnia magna* (Hazardous substances in water towards *Daphnia magna*). Z Wasser Abwasserf 10: 161–166.

Bro-Rasmussen F, Calow P, Canton JH, Chambers PL, Silva Fernandes A, Hoffmann L, Jouany J-M, Klein W, Persoone G, Scoullos M, Tarazona JV, Vighi M (1994) EEC water quality objectives for chemicals dangerous to aquatic environments (List 1). Rev Environ Contam Toxicol 137: 83–110.

Brown MD, Carter J, Thomas D, Purdie DM, Kay BH (2002) Pulse-exposure effects of selected insecticides to juvenile Australian crimson-spotted rainbowfish (*Melanotaenia duboulayi*). J Econ Entomol 95: 294–298.

Bruce RD, Versteeg DJ (1992) A statistical procedure for modeling continuous toxicity data. Environ Toxicol Chem 11: 1485–1494.

Burgess RM, Pelletier MC, Gundersen JL, Perron MM, Ryba SA (2005) Effects of different forms of organic carbon on the partitioning and bioavailability of nonylphenol. Environ Toxicol Chem 24: 1609–1617.

Burr IW (1942) Cumulative frequency functions. Ann Math Stat 13: 215–232.

Burreau S, Axelman J, Broman D, Jakobsson E (1997) Dietary uptake in pike (*Esox lucius*) of some polychlorinated biphenyls, polychlorinated naphthalenes and polybrominated diphenyl ethers administered in natural diet. Environ Toxicol Chem 16: 2508–2513.

Calabrese EJ, Baldwin LA (1993) Performing ecological risk assessments. Lewis Publishers, Chelsea, MI, pp 170–171.

California DPR (2005a) Registration Desk Manual, Chapter 6, Data requirements for obtaining product registration and for label amendments. California Department of Pesticide Regulation, Sacramento, CA.

California DPR (2005b) Pesticide Use Report, http://www.cdpr.ca.gov/docs/pur/pur03rep/03_pur.htm, California Department of Pesticide Regulation, Sacramento, CA.

California SWRCB (2005) State Water Resources Control Board web site. http://www.swrcb.ca.gov/about/mission.html.

Callaghan A, Fisher TC, Grosso A, Holloway GJ, Crane M (2002) Effect of temperature and pirimiphos methyl on biochemical biomarkers in *Chironomus riparius* Meigen. Ecotox Environ Saf 52: 128–133.

Campbell E, Palmer MJ, Shao Q, Warne MStJ, Wilson D (2000) BurrliOZ: A computer program for calculating toxicant trigger values for the ANZECC and ARMCANZ water quality guidelines. Perth, Western Australia.

CCME (1997) Protocol for the derivation of Canadian tissue residue guidelines for the protection of wildlife that consume aquatic biota. Canadian Council of Ministers of the Environment, Ottawa.

CCME (1999) A protocol for the derivation of water quality guidelines for the protection of aquatic life. Canadian Environmental Quality Guidelines. Canadian Council of Ministers of the Environment, Ottawa.

Chapman PM, Fairbrother A, Brown D (1998) A critical evaluation of safety (uncertainty) factors for ecological risk assessment. Environ Toxicol Chem 17: 99–108.

Cold A, Forbes VE (2004) Consequences of a short pulse of pesticide exposure for survival and reproduction of *Gammarus pulex*. Aquat Toxicol 67: 287–299.

Cox C (1987) Threshold dose-response models in toxicology. Biometrics 43: 511–524.

Crane M (1997) Research needs for predictive multispecies tests in aquatic toxicology. Hydrobiologia 346: 149–155.

Crane M, Attwood C, Sheahan D, Morris S (1999) Toxicity and bioavailability of the organophosphorous insecticide pirimiphos methyl to the freshwater amphipod *Gammarus pulex* L. in laboratory and mesocosm systems. Environ Toxicol Chem 18: 1456–1461.

Crane M, Chapman PF, Sparks T, Fenlon J, Newman MC (2002) Can risk assessment be improved with time to event models? In: Risk Assessment with Time to Event Models, Crane M, Newman MC, Chapman PF, Fenlon J (eds), Lewis Publishers, Boca Raton, FL, pp 153–166.

Crane M, Newman MC (2000) What level of effect is a no observed level? Environ Toxicol Chem 19: 516–519.

Crane M, Sildanchandra W, Kheir R, Callaghan A (2002) Relationship between biomarker activity and developmental endpoints in *Chironomus riparius* Meigen exposed to an organophosphate insecticide. Ecotox Environ Saf 53: 361–369.

Cronin MTD, Walker JD, Jaworska JS, Comber MHI, Watts CD, Worth AP (2003) Use of QSARs in international decision-making frameworks to predict ecologic effects and environmental fate of chemical substances. Environ Health Perspect 111: 1376–1390.

CSIRO (2001) BurrliOZ v. 1.0.13. Commonwealth Scientific and Industrial Research Organization, Australia.

CSTE/EEC (1987) Internal report CSTE/87/101/XI from Directorate/EEC General for Environment, Nuclear Safety and Civil Protection, DG XI/A/2, Brussels.

Curtis H, Barnes SN (1981) Invitation to biology, 3rd Ed. Worth Publishers, Inc. New York.

CVRWQCB (2004) Water Quality Control Plan for the Sacramento River and San Joaquin River Basins, 4th Edition (as amended in 2004). Central Valley Regional Water Quality Control Board, Rancho Cordova, CA.

Daily GC, Ehrlich PR, Haddad NM (1993) Double keystone bird in a keystone species complex. Proc Natl Acad Sci USA 90: 592–594.

Daniels RE, Allan JD (1981) Life table evaluation of chronic exposure to a pesticide. Can J Fish Aquat Sci 38: 485–494.

De Coen WM, Janssen CR (2003) A multivariate biomarker-based model predicting population-level responses of *Daphnia magna*. Environ Toxicol Chem 22: 2195–2201.

Del Carmen Alvarez M, Fuiman LA (2005) Environmental levels of atrazine and its degradation products impair survival skills and growth of red drum larvae. Aquat Toxicol 74: 229–241.

Dileanis PD, Bennett KP, Domagalski JL (2002) Occurrence and transport of diazinon in the Sacramento River, California, and selected tributaries during three winter storms, January-February, 2000. United States Geological Survey, Water-Resources Investigations Report 02–4101.

Dileanis PD, Brown DL, Knifong DL, Saleh D (2003) Occurrence and transport of diazinon in the Sacramento River and selected tributaries, California, during two winter storms, January-February, 2001. United States Geological Survey, Water-Resources Investigations Report 03–4111.

Di Toro DM (2003) Executive Summary. In: Re-evaluation of the State of the Science for Water-Quality Criteria Development. SETAC Press, Pensacola, FL, pp xxi–xxv.

Dixon PM, Newman MC (1991) Analyzing toxicity data using statistical models of time-to-death: An introduction. In: Newman MC, McIntosh AW (eds), Metal Ecotoxicology: Concepts and Applications. Lewis Publishers, Inc., Chelsea, MI., pp 207–242

Domagalski J (2000) Pesticides in surface water measured at select sites in the Sacramento River basin, California 1996–1998. United States Geological Survey, Water-Resources Investigations Report 00–4203.
Duboudin C, Ciffroy P, Magaud H (2004) Acute-to-chronic species sensitivity distribution extrapolation. Environ Toxicol Chem 23: 1774–1785.
Dubrovsky NM, Kratzer CR, Brown LR, Gronberg JAM, Burow KR (1998) Water quality in the San Joaquin-Tulare Basin, California 1992–95, United States Geological Survey Circular 1159, on line @ URL:http://water.usgs.gov/pubs/circ1159, updated April 20, 1998.
ECB (2003) Technical guidance document on risk assessment in support of commission directive 93/67/EEC on risk assessment for new notified substances, commission regulation (EC) no. 1488/94 on risk assessment for existing substances, directive 98/8/EC of the European Parliament and of the Council concerning the placing of biocidal products on the market. Part II. Environmental Risk Assessment. European Chemicals Bureau, European Commission Joint Research Center, European Communities.
ECETOC (1993) Aquatic toxicity data evaluation. ECETOC Technical Report No. 56. ECETOC, Brussels.
ECOFRAM (1999) Committee on FFRA risk assessment methods aquatic report. US Environmental Protection Agency, Washington, DC.
Egeler P, Meller M, Roembke J, Spoerlein P, Streit B, Nagel R (2001) *Tubifex tubifex* as a link in food chain transfer of hexachlorobenzene from contaminated sediment to fish. Hydrobiologia 463: 171–184.
Emans HJB, Van Den Plassche EJ, Canton JH (1993) Validation of some extrapolation methods used for effect assessment. Environ Toxicol Chem 12: 2139–2154.
Erickson RJ, Stephan CE (1988) Calculation of the final acute value for water quality criteria for aquatic organisms. EPA/600/3-88-018. US Environmental Protection Agency, Washington, DC.
Ericksson L, Jaworska J, worth AP, Cronin MTD, McDowell RM, Gramatica P (2003) Methods for reliability and uncertainty assessment and for applicability evaluations of classification- and regression-based QSARs. Environ Health Perspect 111: 1361–1375.
EU (2000) Council Directive of 23 October 2002. Establishing a framework for community action in the field of water policy (2000/60/EC). Official Journal of the European Communities, L327, 22 December.
EVS (1999) A critique of the ANZECC and ARMCANZ (1999) water quality guidelines. Prepared for: Minerals Council of Australia and Kwinana Industries Council. Final Report, October 1999, EVS, Vancouver, BC.
Felsot AS (2005) A critical analysis of the draft report, "Amendments to the Water Quality Control Plan for the Sacramento River and San Joaquin River Basins for the Control of Diazinon and Chlorpyrifos Runoff into the Lower San Joaquin River" (Karkoski et al. 2004) and supporting documents. Prepared for the Central Valley Regional Water Quality Control Board, Sacramento, CA.
Fisher DJ, Burton DT (2003) Comparison of two US Environmental Protection Agency species sensitivity distribution methods for calculation ecological risk criteria. Hum Ecol Risk Assess 9: 675–690.
Fisk AT, Norstrom RF, Cymbalisty CD, Muir DCG (1998) Dietary accumulation and depuration of hydrophobic organochlorines: Bioaccumulation parameters and their relationship with the octanol/water partition coefficient. Environ Toxicol Chem 17: 951–961.
Forbes VE, Cold A (2005) Effects of the pyrethroid esfenvalerate on life-cycle traits and population dynamics of *Chironomus riparius*—Importance of exposure scenario. Environ Toxicol Chem 24: 78–86.
Fox DR (1999) Setting water quality guidelines—A statistician's perspective. SETAC News 19(3): 17–18.
Gentile H, Gentile SM, Hairston HG Jr, Sullivan BK (1982) The use of life-tables for evaluating the chronic toxicity of pollutants to *Mysidopsis bahia*. Hydrobiologia 93: 179–187.
GESAMP (1989) The evaluation of the hazards of harmful substances carried by ships: Revision of GESAMP reports and studies No. 17. IMO Reports and studies No. 35. Group of Experts on the Scientific Aspects of Marine Protection (United Nations).

GLEC (2003) Draft compilation of existing guidance for the development of site-specific water quality objectives in the state of California. Great Lakes Environmental Center, Columbus, OH.

Giesy JP, Solomon KR, Coates JR, Dixon KR, Giddings JF, Kenaga EE (1999) Chlorpyrifos: Ecological risk assessment in North American aquatic environments. Rev Environ Contam Toxicol 160: 1–129.

Gobas FAPC, Muir DCG, Mackay D (1988) Dynamics of dietary bioaccumulations and faecal elimination of hydrophobic chemicals in fish. Chemosphere 17: 943–962.

Government of British Columbia (1995) Derivation of water quality criteria to protect aquatic life in British Columbia. Government of British Columbia, Ministry of Land, Air and Water Protection, Water Quality Branch, http://wlapwww.gov.bc.ca/wat/wq/BCguidelines/derive.html#can.

Grist EPM, Crane M, Jones C, Whitehouse P (2003) Estimation of demographic toxicity through the double bootstrap. Wat Res 37: 618–626.

Grist EPM, Leung KMY, Sheeler JR, Crane M (2002) Better bootstrap estimation of hazardous concentration thresholds for aquatic assemblages. Environ Toxicol Chem 21: 1515–1524.

Grothe DR, Kickson KL, Reed-Judkins DK (eds) (1996) Whole effluent toxicity testing: An evaluation of methods and prediction of receiving system impacts. SETAC Press, Pensacola, FL.

Hansch C, Leo A (1979) Substituent constants for correlation analyses in chemistry and biology. John Wiley and Sons, New York, NY.

Hansch C, Leo A, Hoekman D (1995) Exploring QSAR. Hydrophobic, electronic, and steric constants. American Chemical Society, Washington, DC.

Hanson ML, Sanderson H, Solomon KR (2003) Variation, replication, and power analysis of *Myriophyllum* spp. microcosm toxicity data. Environ Toxicol Chem 22: 1318–1329.

Heckman L-H, Friberg N (2005) Macroinvertebrate community response to pulse exposures with the insecticide lambda-cyhalothrin using in-stream mesocosms. Environ Toxicol Chem 24: 582–590.

Hodson PV, Blunt BR, Borgmann U, Minns CK, McGraw S (1983) Effect of fluctuating lead exposures on lead accumulation by rainbow trout (*Salmo gairdneri*). Environ Toxicol Chem 2: 225–238.

Hoekstra JA, Van Ewijk (1993) Alternatives for the no-observed-effect level. Environ Toxicol Chem 12: 187–194.

Hose GC, Van Den Brink PJ (2004) Confirming the species-sensitivity distribution concept for endosulfan using laboratory, mesocosm, and field data. Arch Environ Contam Toxicol 47: 511–520.

Host GE, Regal RR, Stephan CE (1995) Analyses of acute and chronic data for aquatic life. US. Environmental Protection Agency, Washington, DC.

Howard PH (1990) Handbook of environmental fate and exposure data for organic chemicals. Vol. II: Solvents. ISBN 0–87371–204–8, Lewis Publishers, Chelsea, MI.

Howard PH (1991) Handbook of environmental fate and exposure data for organic chemicals. Vol. III: Pesticides. ISBN 0–83731–328–1, Lewis Publishers, Chelsea, MI.

Ingersoll CG, Winner RW (1982) Effect on *Daphnia pulex* (de geer) of daily pulse exposures to copper or cadmium. Environ Toxicol Chem 1: 321–327.

Irmer U, Markard C, Blondzik K, Gottschalk C, Kussatz C, Rechenberg B, Schudoma D (1995) Quality targets for concentrations of hazardous substances in surface waters in Germany. Ecotox Environ Saf 32: 233–243.

Jagoe RH, Newman MC (1997) Bootstrap estimation of community NOEC values. Ecotoxicol 6: 293–306.

Jaworska JS, Comber M, Auer C, Van Leeuwen CJ (2003) Summary of a workshop on regulatory acceptance of (Q)SARs for human health and environmental endpoints. Environ Health Perspectives 111: 1358–1360.

Kem I (1989) Environmentally hazardous substances. List of examples and scientific documentation. Kemikalieinspektionen 10/89, 303 pp (in Swedish).

Kenaga EE (1982) Predictability of chronic toxicity from acute toxicity of chemicals in fish and aquatic invertebrates. Environ Toxicol Chem 1: 347–358.

Kloepper-Sams PJ, Owens JW (1993) Environmental biomarkers as indicators of chemical exposure. J Haz Mat 35: 283–294.
Könemann H (1981) Fish toxicity tests with mixtures of more than two chemicals: A proposal for a quantitative approach and experimental results. Toxicology 19: 229–238.
Kooijman SALM (1987) A safety factor for LC50 values allowing for differences in sensitivity among species. Wat Res 21: 269–276.
Kooijman SALM, Bedaux JJM (1996a) Analysis of toxicity tests on fish growth. Wat Res 30: 1633–1644.
Kooijman SALM, Bedaux JJM (1996b) Analysis of toxicity tests on *Daphnia* survival and reproduction. Wat Res 30: 1711–1723.
Kooijman SALM (1993) Dynamic Energy Budgets in biological systems. Theory and applications in ecotoxicology. Cambridge University Press, Cambridge, UK.
Kooijman SALM, Hanstveit AO, Nyholm N (1996) No-effect concentrations in algal growth inhibition tests. Wat Res 30: 1625–1632.
Kratzer CR, Zamora C, Knifong DL (2002) Diazinon and chlorpyrifos loads in the San Joaquin River Basin, California. January and February 2000. United States Geological Survey, Water-Resources Investigations Report 02–4103.
Kraufvelin P (1999) Baltic hard bottom mesocosms unplugged: Replicability, repeatability and ecological realism examined by non-parametric multivariate techniques. J Exp Mar Biol Ecol 240: 229–258.
Kuivila KM, Barnett HD, Edmunds JL (1999) Herbicide concentrations in the Sacramento-San Joaquin Delta, California. United States Geological Survey, Water-Resources Investigations Report 99–4018 B.
La Point TW, Belanger SE, Crommentuijn T, Goodrich-Mahoney J, Kent RA, Mount DI, Spry DJ, Vigerstad T, Di toro DM, Keating FJ Jr, Reiley MC (2003) Problem Formulation. In: Reevaluation of the State of the Science for Water-Quality Criteria Development. SETAC Press, Pensacola, FL, pp 1–14.
LAWA (1997) Zielvorgaben sum Schutz oberirdischer Binnengewässer. Band 1. Länderarbeitsgemeinschaft Wasser. Kulturbuchverlag Berlin GmbH, Berlin.
Lawton JH (1994) What do species do in ecosystems? Oikos 71: 367–374.
Lemly AD (1985) Toxicology of selenium in a freshwater reservoir: Implications for environmental hazard evaluation and safety. Ecotox Environ Saf 10: 314–348.
Lepper P (2002) Towards the derivation of quality standards for priority substances in the context of the Water Framework Directive. Final Report of the Study Contract No. B4–3040/2000/30673/MAR/E1. Fraunhofer-Institute Molecular biology and Applied Ecology, Munich.
Lillebo HP, Shaner S, Carlson D, Richard N (1988) Water quality criteria for selenium and other trace elements for protection of aquatic life and its uses in the San Joaquin Valley. In: Technical Committee Report: Regulation of agricultural drainage to the San Joaquin River. Appendix D. California State Water Resources Control Board, Sacramento, CA.
LOGKOW (1994) LOGKOW octanol-water partition coefficient program. Now called KowWin. Available at http://www.syrres.com/esc/est_soft.htm. Syracuse Research Corporation, New York, NY.
Loonen H, Parsons JR, Govres HAJ (1991) Dietary accumulation of PCDDs and PCDFs in guppies. Chemosphere 23: 1349–1357.
Lydy M, Belden J, Wheelock C, Hammock B, Denton D (2004) Challenges in regulating pesticide mixtures. Ecol Soc 9: 1 [online] URL: http://www.ecologyandsociety.org/vol9/iss6/art1/.
Maboeta MS, Reinecke SA, Reinecke AJ (2003) Linking lysosomal biomarker and population responses in a field population of *Aporrectodea caliginosa* (Oligochaaeta) exposed to the fungicide copper oxychloride. Ecotox Environ Saf 56: 411–418.
Mackay D, Shiu W-Y, Ma K-C (1992) Illustrated handbook of physical-chemical properties and environmental fate for organic chemical. Volume I. Monoaromatic hydrocarbons, chlorobenzenes, and PCBs. Lewis Publishers, Boca Raton, USA.
Mackay D, Shiu W-Y, Ma K-C (1993) Illustrated handbook of physical-chemical properties and environmental fate for organic chemical. Volume III. Volatile organic chemicals. Lewis Publishers, Boca Raton, USA.

Mackay D, Shiu W-Y, Ma K-C (1995) Illustrated handbook of physical-chemical properties and environmental fate for organic chemical. Volume IV. Oxygen, nitrogen, and sulfur containing compounds. Lewis Publishers, Boca Raton, USA.

Mackay D, Shiu W-Y, Ma K-C (1997) Illustrated handbook of physical-chemical properties and environmental fate for organic chemical. Volume V. Pesticide chemicals. Lewis Publishers, Boca Raton, USA.

Mackay D, Shiu W-Y, Ma K-C (1999) Illustrated handbook of physical-chemical properties and environmental fate for organic chemical. CRC-LLC netbase, CD-rom version.

Maltby L, Blake N, Brock TCM, Van Den Brink PJ (2005) Insecticide species sensitivity distributions: Importance of test species and relevance to aquatic ecosystems. Environ Toxicol Chem 24: 379–388.

Matthiessen P (2000) Is endocrine disruption a significant ecological issue? Ecotoxicol 9: 21–24.

Mayer FL, Krause GF, Buckler DR, Ellersieck MR, Lee G (1994) Predicting chronic lethality of chemicals to fishes from acute toxicity test data: Concepts and linear regression analysis. Environ Toxicol Chem 13: 671–678.

Mayer FL, Ellersieck MR, Krause GF, Sun K, Lee G, Buckler DR (2002) Time-concentration-effect models in predicting chronic toxicity from acute toxicity data. In: Risk Assessment with Time to Event Models, Crane M, Newman MC, Chapman PF, Fenlon J (eds), Lewis Publishers, Boca Raton, FL, pp 39–67.

Menconi M, Beckman J (1996) Hazard assessment of the insecticide methomyl to aquatic organisms in the San Joaquin river system. Admin. Rep. 96–6. California Department of Fish and Game, Environ. Serv. Div., Rancho Cordova, CA.

Mensink BJWG, Montforts M, Wijkhuizen-Maslankiewicz L, Tibosch H, Linders JBHJ (1995) Manual for summarizing and evaluating the environmental aspects of pesticides. Report no. 67101022. National Institute of Public Health and Environmental Protection (RIVM), Bilthoven, The Netherlands.

MHSPE (1994) Environmental Quality Objectives in the Netherlands: A Review of Environmental Quality Objectives and Their Policy Framework in the Netherlands. Ministry of Housing, Spatial Planning and the Environment, The Hague.

MITI (1992) Biodegradation and bioaccumulation data on existing data based on the CSCL Japan. Japan Chemical Industry, Ecology-Toxicology and Information Center, Ministry of International Trade and Industry. ISBN 4-89074-101-1.

Moore DRJ, Caux PY (1997) Estimating low toxic effects. Environ Toxicol Chem 16: 794–801.

Mount DR, Ankley GT, Brix KV, Clements WH, Dixon DG, Fairbrother A, Hickey CW, Lanno RP, Lee CM, Munns WR, Ringer RK, Staveley JP, Wood CM, Erickson RJ, Hodson PV (2003) Effects assessment: Introduction. In: Reevaluation of the State of the Science for Water-Quality Criteria Development, Reiley MC, Stubblefield WA, Adams WJ, Di Toro DM, Hodson PV, Erickson RJ, Keating FJ Jr (eds), SETAC Press, Pensacola, FL.

Murray FJ, Smith FA, Nitschke KD, Humiston CG, Kociba RJ, Schwetz BA (1979) Three generation reproduction study of rats given 2,3,7,8-tetrachlorodibenzo-p-dioxin (TCDD) in the diet. Toxicol Appl Pharmacol 50: 241–252.

Nabholz JV (1991) Environmental hazard and risk assessment under the United States Toxic Substances Control Act. Sci Total Environ 109/110: 649–665.

Nabholz JV (2003) Toxicity assessment, risk assessment, and risk management of chemicals under TSCA in USA. Office of Pollution Prevention and Toxics, United States Environmental Protection Agency, Washington, DC.

Newman MC, Aplin M (1992) Enhancing toxicity data interpretation and prediction of ecological risk with survival time modeling: An illustration using sodium chloride toxicity to mosquitofish (Gambusia holbrooki). Aquatic Toxicol 23: 85–96.

Newman MC, McCloskey JT (1996) Time-to-event analysis of ecotoxicity data. Ecotoxicology 5: 187–196.

Newman MC, Crane M (2002) Introduction to time to event methods. In: Risk Assessment with Time to Event Models, Crane M, Newman MC, Chapman PF, Fenlon J (eds), Lewis Publishers, Boca Raton, FL, pp 1–6.

Newman MC, Ownby DR, Mézin LCA, Powell DC, Christensen TRL, Lerberg SB, Anderson B-A (2000) Applying species-sensitivity distributions in ecological risk assessment: Assumptions of distribution type and sufficient numbers of species. Ecotoxicol Environ Chem 19: 508–515.

Newman MC, Ownby DR, Mézin LCA, Powell DC, Christensen TRL, Lerberg SB, Anderson B-A, Padma TV (2002) Species sensitivity distributions in ecological risk assessment: Distributional assumptions, alternate bootstrap techniques, and estimation of adequate number of species. In: Species Sensitivity Distributions in Ecotoxicology, Posthuma L, Suter II GW, Traas TP (eds), Lewis Publishers, CRC Press, New York, NY.

Nikunen E, Leinonen R, Kultamaa A (1990) Environmental properties of chemicals. Ministry of the Environment, Research report 91, Finland.

Norstrom RJ, McKinnon AE, De Freitas ASW (1976) A bioenergetics-based model for pollutant accumulations by fish. Simulation of PCB and methylmercury in Ottawa River yellow perch *Perca flavescens*. J Fish Res Board Can 33: 248–267.

North Carolina Department of Environment and Natural Resources (2003) "Redbook" Surface Waters and Wetlands Standards. North Carolina DENR, Division of Water Quality, NC Administrative Code 15A NCAC 02 B.0100 and.0200.

Novartis Crop Protection (1997) An ecological risk assessment of diazinon in the Sacramento and San Joaquin River basins. Novartis Crop Protection, Environmental and Public Affairs Department Technical Report 11/97, Greensboro, NC.

OECD (1992) Fish, early-life stage toxicity test. OECD guidelines for testing of chemicals. Organization for Economic Co-operation and Development, Paris.

OECD (1995) Guidance Document for Aquatic Effects Assessment. Organization for Economic Co-operation and Development, Paris.

Okkerman PC, Van Den Plassche EJ, Emans HJB, Canton JH (1993) Validation of some extrapolation methods with toxicity data derived from multiple species experiments. Ecotox Environ Saf 25: 341–359.

Okkerman PC, Van Den Plassche EJ, Slooff W, Van Leeuwen CJ, Canton JH (1991) Ecotoxicological effects assessment: A comparison of several extrapolation procedures. Ecotox Environ Saf 21: 182–193.

Olsen T, Elerbeck L, Fisher T, Callaghan A, Crane M (2001) Variability in acetylcholinesterase and glutathione *S*-transferase activities in *Chironomus riparius* Meigen deployed *in situ* at uncontaminated field sites. Environ Toxicol Chem 20: 1725–1732.

Parkhurst DF (1998) Arithmetic versus geometric means for environmental concentration data. Environ Sci Technol 32: 92A–98A.

Pawlisz AV, Busnarda J, McLauchlin A, Caux P-Y, Kent RA (1998) Canadian water quality guidelines for deltamethrin. Environ Toxic Water 13: 175–210.

Persoone G, Janssen CR (1994) Field validation of predictions based on laboratory toxicity tests. In: Freshwater Field Tests for Hazard Assessment of Chemicals. Hill IR, Heimbach F, Leeuwangh P, Matthiessen P (eds), CRC Press, Boca Raton, FL, pp 379–397.

Péry ARR, Flammarion P, Vollat B, Bedaux JJM, Kooijman SALM, Garric J (2002) Using a biology-based model (DEBtox) to analyze bioassays in ecotoxicology: Opportunities and recommendations. Environ Toxicol Chem 21: 459–465.

Plackett RL, Hewlett PS (1952) Quantal responses to mixtures of poisons. J Royal Stat Soc B 14: 141–163.

Posthuma L, Traas TP, Suter GW III (2002a) General introduction to species sensitivity distributions. In: Species Sensitivity Distributions in Ecotoxicology, Posthuma L, Suter GW, Traas TP IIII (eds), Lewis Publishers, CRC Press, Boca Raton, FL, pp 3–10.

Posthuma L, Traas TP, De Zwart D, Suter GW II (2002b) Conceptual and technical outlook on species sensitivity distributions. In: Species Sensitivity Distributions in Ecotoxicology, Posthuma L, Suter GW, Traas TP II(eds), Lewis Publishers, CRC Press, Boca Raton, FL, pp 475–508.

Pusey BJ, Arthington AH, McClean J (1994) The effects of a pulsed application of chlorpyrifos on macroinvertebrate communities in an outdoor artificial stream system. Ecotox Environ Saf 27: 221–250.

Qiao P, Gobas FAPC, Farrell AP (2000) Relative contributions of aqueous and dietary uptake of hydrophobic chemicals to the body burden in juvenile rainbow trout. Environ Contam Toxicol 39: 369–377.

Ramos EU, Vaes WHJ, Verhaar HJM, Hermens JLM (1998) Quantitative structure-activity relationships for the aquatic toxicity of polar and nonpolar narcotic pollutants. J Chem Informat Comput Sci 38: 845–852.

Reiley MC, Stubblefield WA, Adams WJ, Di Toro DM, Hodson PV, Erickson RJ, Keating FJ Jr (2003) Reevaluation of the state of the science for water-quality criteria development. SETAC Press, Pensacola, FL.

Reynaldi S, Liess M (2005) Influence of duration of exposure to the pyrethroid fenvalerate on sublethal responses and recovery of *Daphnia magna* Straus. Environ Toxicol Chem 24: 1160–1164.

Rider CV, LeBlanc GA (2005) An integrated addition and interaction model for assessing toxicity of chemical mixtures. Toxicol Sci 87: 520–528.

Rio Convention (1992) United Nations Conference on Environment and Development: Rio Declaration on Environment and Development, June 14, 1992. Reprinted in Intl. Legal Materials 31: 874–879.

RIVM (2001) Guidance document on deriving environmental risk limits in The Netherlands. Report no. 601501 012. Traas TP(ed), National Institute of Public Health and the Environment, Bilthoven, The Netherlands.

RIVM (2004) ETX 2.0. Normal distribution based hazardous concentration and fraction affected. Designed by Van Vlaardingen P, Traas T, Aldenberg T, Wintersen A. National Institute of Public Health and the Environment, Bilthoven, The Netherlands.

Romijn CAFM, Luttik R, Van De Meent D, Slooff W, Canton JH (1993) Presentation of a general algorithm to include effect assessment on secondary poisoning in the derivation of environmental quality criteria. Ecotox Environ Saf 26: 61–85.

Roth L (1993) Wassergefärdende Stoffe. Ecomed verlag Gmbh, Landsberg/Lech.

Roux DJ, Jooste SHJ, MacKay HM (1996) Substance-specific water quality criteria for the protection of South African freshwater ecosystems: Methods for derivation and initial results for some inorganic toxic substances. S African J Sci 92: 198–206.

Russom CL, Bradbury SP, Broderius SJ, Hammermeister DE, Drummond RA (1997) Predicting modes of toxic action from chemical structure: Acute toxicity in the fathead minnow (*Pimephales promelas*). Environ Toxicol Chem 16: 948–967.

Samsoe-Petersen L, Pedersen F (eds) (1995) Water quality criteria for selected priority substances, Working Report, TI 44. Water Quality Institute, Danish Environmental Protection Agency, Copenhagen, Denmark.

Sanderson H (2002) Pesticide studies—Replication of micro/mesocosm studies. Environ Sci Pollut R 6: 429–435.

Schwarzenbach RP, Gschwend PM, Imboden DM (1993) Environmental organic chemistry. John Wiley and Sons, Inc., NY, USA

Schulz R, Liess M (2001) Toxicity of fenvalerate to caddisfly larvae: Chronic effects of 1- vs. 10-h pulse-exposure with constant exposures. Chemosphere 41: 1511–1517.

Segner H (2005) Developmental, reproductive, and demographic alterations in aquatic wildlife: Establishing causality between exposure to endocrine-active compounds (EACs) and effects. Acta Hudrochim Hydrobiol 33: 17–26.

SETAC-Europe (1992) Guidance document on testing procedures for pesticides in freshwater mesocosms. From a meeting of experts on guidelines for static field mesocosm tests, held at Monks Wood Experimental Station, Abbotts Ripton, Huntingdon, UK, 3–4 July 1991.

Shao Q (2000) Estimation for hazardous concentrations based on NOEC toxicity data: An alternative approach. Envirometrics 11: 583–595.

Siepmann S, Jones MR (1998) Hazard assessment of the insecticide carbaryl to aquatic organisms in the Sacramento-San Joaquin river system. Admin. Rep. 98–1. California Department of Fish and Game, Office of Spill Prevention and Response, Rancho Cordova, CA.

Slooff W (1992) RIVM guidance document. Ecotoxicological effect assessment: Deriving maximum tolerable concentrations (MTC) from single-species toxicity data. Report 719102 018, RIVM Bilthoven, The Netherlands.

Solomon KR, Giddings JM, Maund SJ (2001) Probabilistic risk assessment of cotton pyrethroids: I. Distributional analyses of laboratory aquatic toxicity data. Environ Toxicol Chem 20: 652–659.

Solomon KR, Takacs P (2002) Probabilistic risk assessment using species sensitivity distributions. In: Species Sensitivity Distributions in Ecotoxicology, Posthuma L, Suter GW II, Traas TP(eds), Lewis Publishers, New York, NY, pp 285–314.

Speijers GJA, Franken MAM, Van Leeuwen FXR, Van Egmond HP, Boot R, Loeber JG (1986) Subchronic oral toxicity study of patulin in the rat. Report 618314001. RIVM Bilthoven, The Netherlands.

Spromberg JA, Birge WJ (2005a) Modeling the effects of chronic toxicity on fish populations: The influence of life-history strategies. Environ Toxicol Chem 24: 1532–1540.

Spromberg JA, Birge WJ (2005b) Population survivorship index for fish and amphibians: Application to criterion development and risk assessment. Environ Toxicol Chem 24: 1541–1547.

Stephan CE (1985) Are the "Guidelines for deriving numerical national water quality criteria for the protection of aquatic life and its uses" based on sound judgments? In: Aquatic Toxicology and Hazard Assessment: Seventh Symposium, ASTM STP 854, Cardwell RD, Purdy R, Bahner RC (eds), American Society for Testing and Materials, Philadelphia, PA, pp 515–526.

Stephan CE, Rogers JW (1985) Advantages of using regression analysis to calculate results of chronic toxicity tests. In: Aquatic Toxicology and Hazard Assessment: Eighth Symposium. ASTM STP 891, Bahner RC, Hansen DJ (eds), American Society for Testing and Materials, Philadelphia, PA, pp 328–338.

Sun K, Krause GJ, Mayer FL Jr, Ellersieck MR, Basu AP (1995) Predicting chronic lethality of chemicals to fishes from acute toxicity test data: Theory of accelerated life testing. Environ Toxicol Chem 14: 1745–1752.

Suter GW II (2002) North American history of species sensitivity distributions. In: Species Sensitivity Distributions in Ecotoxicology, Posthuma L, Suter GW II, Traas TP (eds), Lewis Publishers, CRC Press, Boca Raton, FL, pp 11–17.

Suter GW II, Rosen AE, Linder E, Parkhurst DF (1987) Endpoints for responses of fish to chronic toxic exposures. Environ Toxicol Chem 6: 793–809.

Teh SJ, Deng DF, Werner I, Teh FC, Hung SSO (2005) Sublethal toxicity of orchard stormwater runoff in Sacramento splittail (*Pogonichthys macrolepidotus*) larvae. Mar Environ Res 59: 203–216.

Thomann RV, Connolly JP (1984) Model of PCB in the Lake Michigan lake trout food chain. Environ Sci Technol 18: 65–71.

Traas TP, Van De Meent D, Posthuma L, Hamers T, Kater BJ, De Zwart D, Aldenberg T (2002) The potentially affected fraction as a measure of ecological risk. In: Species Sensitivity Distributions in Ecotoxicology, Posthuma L, Suter GW II, Traas TP(eds), Lewis Publishers, New York, NY, pp 315–344.

Traas TP, Van Wezel AP, Hermens JLM, Zorn M, Van Hattum AGM, Van Leeuwen CJ (2004) Environmental quality criteria for organic chemicals predicted form internal effect concentrations and a food web model. Environ Toxicol Chem 23: 2518–2527.

Treibskorn R, Adam S, Behrens A, Beier S, Böhmer J, Braunbeck T, Casper H, Dietze U, Gernhöfer M, HOnnen W, Köhler H-R, Körner W, Konradt J, Lehmann R, Luckenbach T, Oberemm A, Schwaiger J, Segner H, Strmac M, Schüürmann G, Siligato S, Traunspurger W (2003) Establishing causality between pollution an defects at different levels of biological organization: The VALIMAR project. Hum Ecol Risk Assess 9: 171–194.

USEPA (1984a) Guidelines for deriving numerical aquatic site-specific water quality criteria by modifying national criteria. EPA-600/3–84–099. US Environmental Protection Agency, Washington, DC.

USEPA (1984b) Estimating "concern levels" for concentrations of chemical substances in the environment. Environmental Effects Branch, Health and Environmental Review Division (TS-

796), Office of Toxic Substances, US Environmental Protection Agency, Washington, DC 20460–0001.
USEPA (1985) Guidelines for deriving numerical national water quality criteria for the protection of aquatic organisms and their uses. PB-85-227049. US Environmental Protection Agency, National Technical Information Service, Springfield, VA, USA.
USEPA (1986) Guidelines for deriving ambient aquatic life advisory concentrations. EPA/822/R86/100. US Environmental Protection Agency, Washington, DC.
USEPA (1987) 40 CFR Part 797. Environmental Effects Testing Guidelines. US Environmental Protection Agency, Washington, DC.
USEPA (1991) Technical Support Document for Water Quality-Based Toxics Control. EPA/505/2-90-001. US Environmental Protection Agency, Washington, DC.
USEPA (1994) Water Quality Standards Handbook. EPA-823-B-94-005. US Environmental Protection Agency, Washington, DC.
USEPA (1993) Federal Register, 40 CFR Part 158.490.
USEPA (2000) Ambient Aquatic Life Water Quality Criteria for Dissolved Oxygen (Saltwater): Cape Cod to Cape Hatteras. EPA-822-R-00–012. US Environmental Protection Agency, Washington, DC.
USEPA (2002a) Draft report on summary of proposed revisions to the aquatic life criteria guidelines. US Environmental Protection Agency, Washington, DC.
USEPA (2002b) Short-term methods for estimating the chronic toxicity of effluents and receiving waters to freshwater organism, 4th edition. EPA-821-R-02–013. US Environmental Protection Agency, Washington, DC.
USEPA (2003a)Water quality guidance for the Great Lakes system. Federal Register, 40 CFR Part 132. US Environmental Protection Agency, Washington, DC.
USEPA (2003b) Ambient aquatic life water quality criteria for tributyltin (TBT) – Final. EPA 822-R-03-031. US Environmental Protection Agency, Washington, DC.
USEPA (2003c) Acute-to-chronic estimation (ACE v 2.0) with time-concentration-effect models, User manual and software. EPA/600/R-03/107. US Environmental Protection Agency, Washington, DC.
USEPA (2003d) Interspecies correlation estimations (ICE) for acute toxicity to aquatic organisms and wildlife. II. User manual and software. EPA/600/R-03/106. US Environmental Protection Agency, Washington, DC.
USEPA (2005) Science Advisory Board Consultation Document, Proposed Revisions to Aquatic Life Guidelines, Water-Based Criteria, Water-Based Criteria Subcommittee, US Environmental Protection Agency, Washington, DC.
USGS (1998) Pesticides in surface and ground water of the United States: Summary of the results of the National Water Quality Assessment Program (NAWQA). US Geological Survey, Washington, DC.
USGS (2005a) Water Resource Data, California Water Year 2004, Volume 4, Northern Central Valley Basins and the Great Basin from Honey Lake Basin to Oregon State Line. US Geological Survey, Sacramento, CA.
USGS (2005b) Water Resources Data, California Water Year 2004, Volume 3, Southern Central Valley Basins and the Great Basin from Walker River to Truckee River. US Geological Survey, Sacramento, CA.
Vaal M, Van Der Wal JT, Hermens J, Hoekstra J, (1997a) Pattern analysis of the variation in the sensitivity of aquatic species to toxicants. Chemosphere 35: 1291–1309.
Vaal M, Van Der Wal JT, Hoekstra J, Hermens J (1997b) Variation in the sensitivity of aquatic species in relation to the classification of environmental pollutants. Chemosphere 35: 1311–1327.
Van De Meent D (1993) SIMPLEOX: A generic multimedia fate evaluation model. Report number 672702 001. National Institute of Public Health and the Environment, Bilthoven, The Netherlands.
Van De Meent D, Aldenberg T, Canton JH, Van Gesteel CAM, Slooff W (1990) Desire for levels, background study for the policy document "Setting environmental quality standards for water and soil." National Institute of Public Health and the Environment, Bilthoven, The Netherlands.

Van Den Brink PJ, Roelsma J, Van Nes EH, Scheffer M, Brock TCM (2002) PERPEST model, a case-based reasoning approach to predict ecological risks of pesticides. Environ Toxicol Chem 21: 2500–2506.

Van Der Hoeven N (2001) Estimating the 5-percentile of the species sensitivity distributions without any assumptions about the distribution. Ecotoxicol 10: 25–34.

Van Der Hoeven N, Noppert F, Leopold A (1997) How to measure no effect. Part I: Towards a new measure of chronic toxicity in ecotoxicology. Introduction and workshop results. Environmetrics 8: 241–248.

Van Der Oost R, Beyer J, Vermeulen NPE (2003) Fish bioaccumulation and biomarkers in environmental risk assessment: A review. Environ Toxicol Pharm 13: 57–149.

Van Leeuwen CJ, Van Der Zandt PTJ, Aldenberg T, Verhaar HJM, Hermens JLM (1992) Application of QSARs, extrapolation and equilibrium partitioning in aquatic effects assessment. I. Narcotic industrial pollutants. Environ Toxicol Chem 11: 267–282.

Van Straalen NM, Denneman CAJ (1989) Ecotoxicological evaluation of soil quality criteria. Ecotox Environ Safe 18: 241–251.

Van Straalen NM, Van Leeuwen CJ (2002) European history of species sensitivity distributions. In: Species Sensitivity Distributions in Ecotoxicology, Posthuma L, Suter GW, II Traas TP (eds), Lewis Publishers, CRC Press, Boca Raton, FL, pp 19–34.

Verhaar HJM, Van Leeuwen CJ, Hermens JLM (1992) Classifying environmental pollutants. 1: Structure-activity relationships for prediction of aquatic toxicity. Chemosphere 25: 471–491.

Verscheuren K (1983) Handbook of environmental data on organic chemicals, 2nd Ed., Van Nostrand Reinhold Co., New York, NY.

Verscheuren K (2001) Handbook of environmental data on organic chemicals, 4th Ed., CD-ROM, Wiley Interscience, New York, NY.

Versteeg DJ, Belanger SE, Carr GJ (1999) Understanding single-species and model ecosystem sensitivity: Data-based comparison. Environ Toxicol Chem 18: 1329–1346.

VROM (1994) Environmental quality objectives in The Netherlands. Ministry of Housing, Spatial Planning and Environment, The Hague, The Netherlands.

Wagner C, Løkke H (1991) Estimation of ecotoxicological protection levels from NOEC toxicity data. Wat Res 25: 1237–1242.

Warmer H, Van Dokkum R (2002) Water pollution control in the Netherlands, Policy and Practice. RIZA report 2002.009.

Warne MSJ, Hawker DW (1995) The number of components in a mixture determines whether synergistic and antagonistic or additive toxicity predominate: The funnel hypothesis. Ecotox Environ Safety 31: 23–28.

Webster's New Collegiate Dictionary (1976) G. & C. Merriam Co., Springfield, MA.

Werner I, Deanovic LA, Connor V, de Vlaming V, Bailey HC, Hinton DE (2000) Insecticide-caused toxicity to *Ceriodaphnia dubia* (cladocera) in the Sacramento-San Joaquin River Delta, California, USA. Environ Toxicol Chem, 19: 215–227.

Wheeler JR, Grist EPM, Leung KMY, Morritt D, Crane M (2002) Species sensitivity distributions: Data and model choices. Marine Pollut Bull 45: 192–202.

Whitehouse P, Crane M, Grist E, O'Hagan A, Sorokin N (2004) Derivation and expression of water quality standards; opportunities and constraints in adopting risk-based approaches in EQS setting. RandD technical Report P2–157/TR. Environment Agency, Rio House, Almondsbury, Bristol.

Wu J, Laird DA (2004) Interactions of chlorpyrifos with colloidal materials in aqueous systems. J Environ Qual 33: 1765–1770.

Zabel TF, Cole S (1999) The derivation of environmental quality standards for the protection of aquatic life in the UK. J CIWEM 13: 436–440.

Zischke JA, Arthur JW, Hermanutz RO, Hedtke SF, Helgen JC (1985) Effects of pentachlorophenol on invertebrates and fish in outdoor experimental channels. Aquat Toxicol 7: 37–58.

Chapter 3
Platinum Group Elements in the Environment: Emissions and Exposure

Aleksandra Dubiella-Jackowska, Żaneta Polkowska, and Jacek Namieśnik

Contents

1 Introduction

The purpose of this chapter is to address the production, applications, and emission sources of the platinum group elements (PGEs). It includes a presentation of the effects these metals have on humans from occupational and environmental exposure. In addition, a brief literature review is presented that addresses the concentrations of PGEs found in environmental samples such as soil, road dust, road tunnel dust, airborne particulate matter, sediments, wastewater, snow, drinks, and samples of flora and fauna.

PGEs are metals that occur in the lithosphere at very low concentrations. Uses of PGEs are rapidly increasing [e.g., for vehicle exhaust catalysts (VECs), electronics, jewelry production, and the pharmaceutical industry]. Their increasing use is result-

A. Dubiella-Jackowska (✉), Ż. Palkowska, and J. Namieśnik
Department of Analytical Chemistry, Chemical Faculty, Gdańsk University of Technology, G. Narutowicza St. 11/12, 80-952 Gdańsk, Poland
E-mail: Dubiella@poczta.fm

D.M. Whitacre (ed.) *Reviews of Environmental Contamination and Toxicology* Volume 199, doi: 10.1007/978-0-387-09808-1_3,

ing in increased emissions and broader environmental distribution of PGEs in some geographical regions. When released into the environment, PGEs accumulate first in airborne particulate matter, road dust, soil, mud, and water; PGEs may then enter organisms and subsequently undergo bioaccumulation. Scientific results from wide-scale investigations support the thesis that concentrations of PGEs, especially platinum, are increasing in tissues and bodily fluids of humans exposed to these metals (Minakata et al. 2006; Petrucci et al. 2005; Rudolph et al. 2005; Nygren and Lundgren 1997; Brook 2006). According to available data, the highest exposed groups (Leœniewska et al. 2001) are individuals who work in refineries, chemical plants, electronics plants, jewelry production, oncological wards (medical personnel), and road maintenance; also highly exposed are women who have breast implants.

In general, metallic forms of PGEs do not enter the biochemical pathways of organisms. However, some platinum salts such as hexachloro platinate and tetrachloro platinate have proven to be allergenic (Balcerzak 1997). The presence of PGEs in a live organism may cause the following effects: asthma, miscarriage, nausea, hair loss, skin diseases, and, in humans, other serious health problems.

Several review papers, in the literature, have addressed topics relevant to PGEs, including their occurrence and applications (Rao and Reddi 2000), their environmental distribution and speciation (Barefoot 1999), and the threats they pose to human health as a result of anthropogenic emissions (Merget and Rosner 2001).

In this chapter, an attempt was made to synthesize the literature that addresses emission sources and the bioavailability of PGEs in the environment.

2 Characteristics of PGEs

PGEs, also called platinum group metals (PGMs) in the international scientific literature, encompass such metals as platinum (Pt), palladium (Pd), rhodium (Rh), ruthenium (Ru), iridium (Ir), and osmium (Os). Average concentrations of these metals in the earth's crust are presented in Table 1.

Table 1 Upper[a] and lower[b] continental crust concentrations of platinum group elements (PGEs)

Element	UCC[a]/LCC[b] concentration (μg kg^{-1})[c]	UCC concentration (μg kg^{-1})[d]
Pt	0.40	0.51
Pd	0.40	0.52
Rh	0.06	–
Ru	0.10	0.21
Ir	0.05	0.022
Os	0.05	0.031

Pt platinum, *Pd* palladium, *Rh* rhodium, *Ru* ruthenium, *Ir* iridium, *Os* osmium
[a]UCC–Upper continental crust
[b]LCC–Lower continental crust
[c]Wedepohl (1995)
[d]Peucker-Ehrenbrink and Jahn (2001)

PGEs occur in nickel (Ni), copper (Cu), and marcasite ore beds (Bradford 1988). They belong to the noble metal group, and their reactivity is low. PGEs occur in nature as primary alloys consisting mainly of Pt. The specific features of these elements, which make them widely applicable, are as follows (Bernardis et al. 2005): high resistance to chemical corrosion over a wide range of temperatures, high melting point, high mechanical resistance, and high plasticity.

3 Production and Application of PGEs

Presently, PGMs are produced from appropriate ores that are mined in South America, Siberia, and Sudbury (Ontario, Canada); their production has been steadily increasing, since 1970, as a result of growing worldwide demand (World Health Organization, WHO 1991). Total world demand in 2000 and 2006 reached 166.7×10^3 and 215.0×10^3 kg for Pt, 260.5×10^3 and 206.1×10^3 kg for Pd, and 25.3×10^3 and 28.6×10^3 kg for Rh, respectively (Johnson Matthey 2007). Figure 1 presents world demand by use category for the yr 2000 and 2006.

PGEs have exceptional catalytic qualities, which renders them particularly useful in industrial catalytic devices. Moreover, these metals have found application in various other sectors of industry, such as (Pyrzyñska 1998; Resano et al. 2007; Brook 2006) chemicals, petrochemicals, electrical, and electronics, glass production, jewelry production, the medical sector (e.g., components of anticancer drugs, medical silicone gels, and elastomers), and in dentistry.

4 Environmental Emission Sources of PGEs

The broad application of PGEs in various industries has increased emissions of these metals to the environment; in this regard, emissions from vehicle catalytic converters and hospital wastewater discharges are particularly significant.

4.1 Emission of PGEs from Vehicle Exhaust Catalysts

Catalytic converters were first fitted to cars in the USA and Japan in the mid-1970s (in response to new emission standards, such as the US Clean Air Act Amendment of 1970), and in Europe at the beginning of the 1980s (Benemann et al. 2005). These catalysts primarily utilize a mix of Pt and Pd, and have the function of catalytically converting carbon monoxide (CO) and hydrocarbons (HCs) to carbon dioxide (CO_2) and water. Because oxidation catalysts have little effect on NO_xs, the new air standards resulted in the development and introduction of "three-way catalysts." Such catalysts simultaneously oxidize CO and HC while reducing NO_xs to nitrogen (Twigg 2007). The most common three-way catalysts fitted to cars in the 1980s contained Pt and Rb in a 5:1 ratio, Rb playing an important role in promoting the reduction of NO_xs.

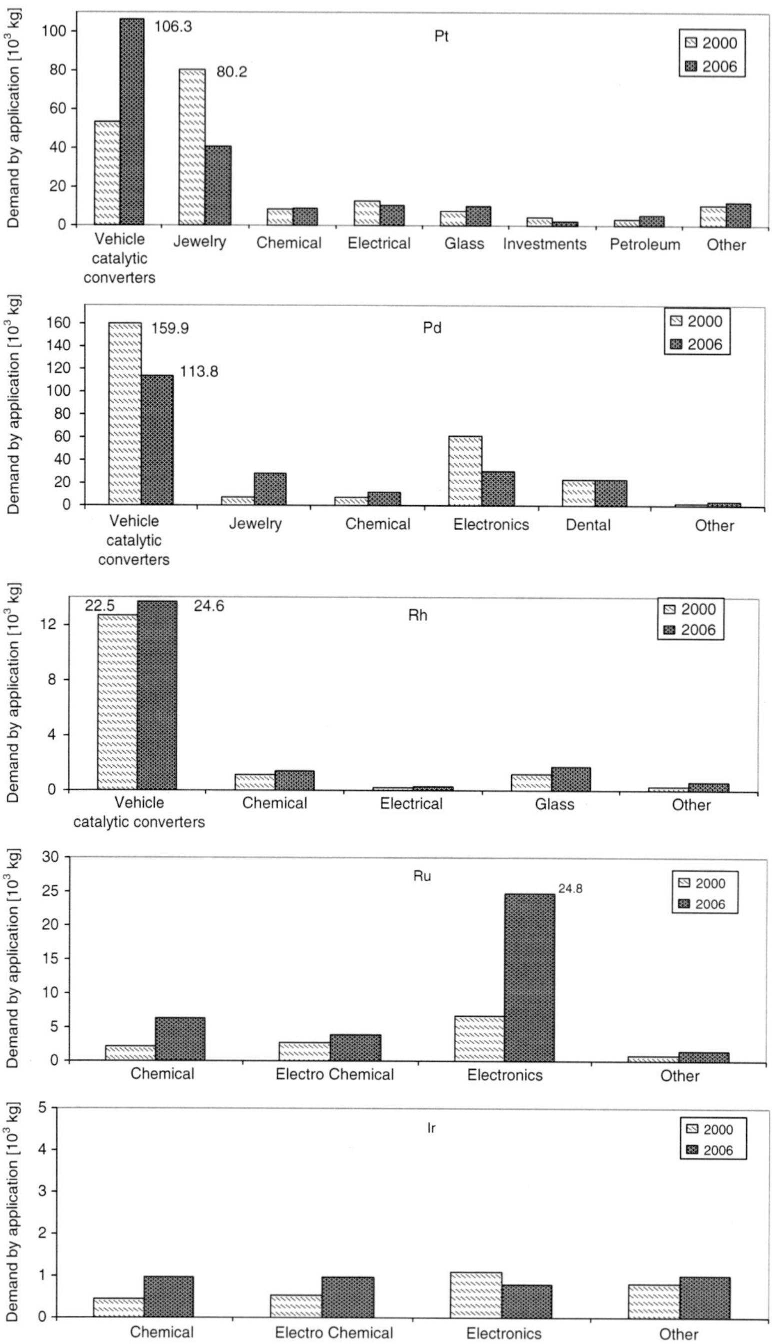

Fig. 1 World consumption of platinum group elements (PGEs) in 2000 and 2006 (Johnson Matthey 2000, 2007). *Pt* platinum, *Pd* palladium, *Rh* rhodium, *Ru* ruthenium, *Ir* iridium

The modern three-way catalysts consist of a ceramic monolith in the form of a honeycomb. A washcoat is applied onto the monolith. The monolith contains small channels, each about 1 mm in diameter (300–600 channels in^{-2}). The washcoat, which includes the active catalyst material, is impregnated on these channel walls. The washcoat consists of porous oxides, such as γ-Al_2O_3, and precious metals (Mitzukami et al. 1991).

Today, there is a wide range of possible combinations and concentrations of Pt, Pd, and Rh required by car manufacturers that are used to achieve different catalyst performance features. In the various Pd–Rh, Pt–Rh, Pt–Pd, Pt-only, or Pt–Pd–Rh catalyst combinations, the percentage of PGEs constitutes 0.1 wt% of the catalysis support material. Other elements present in the catalyst, such as Ce (cerium), Zr (zirconium), and rare earth metals, are used as components of the washcoat to increase specified properties, such as catalyst PGE impregnation, oxygen storage capability, and chemical inertness (Palacios et al. 2000a; Cuif et al. 1996, 1997). The three-way catalytic converters are able to remove more than 90% of CO and NO_xs from exhaust gases, and 80% of unburned HCs (Onovwiona and Uqursal 2006). The automotive CO and NO_x emission-control catalysts used in catalytic converters are usually guaranteed for 40,000–80,000 km, depending on the quality of the product (Shams and Goodarzi 2006).

Moreover, beginning in 1989, an increasing number of diesel passenger cars in Europe were equipped with precious metal-based oxidation catalysts (Van den Tillaard et al. 1996); Pt was and is usually the first choice for these, although recent developments in catalyst research indicate that a partial substitution of Pt by Pd is possible (Heck and Farrauto 2001; Moldovan 2007). These catalysts oxidize gaseous pollutants, such as CO and HCs; in addition, such catalysts oxidize the HC component adsorbed on particulates, thereby reducing the emission of particulates from auto exhaust (Van den Tillaard et al. 1996). The three-way catalytic-converter technology is not applicable to diesel engines because conversions of NO_x to N_2 and CO and of HCs to CO_2 and H_2O will not take place when an excessive amount of air exists. In contrast, oxidation catalysts enable the oxidation of CO and HCs to CO_2 and H_2O, in the presence of excessive oxygen. CO and nonmethane HC conversion levels of 98–99% are achievable, and methane conversion may reach levels of 60–70% (Energy Nexus Group 2002).

When exhaust fumes are released from an engine, chemical and physical reactions occur on the surface of the catalyst carrier as a result of rapidly changing redox conditions, high temperature, and mechanical friction (Bencs et al. 2003). These reactions result in PGE emissions (bound to carrier material) into the environment at ng km^{-1}rates (Moldovan et al. 1999; Ek et al. 2004). Pt, emitted from catalytic converters, exists as surface-oxidized metal nanoparticles, and is bound to larger Al_2O_3 particles (Rühle et al. 1997). Based on the distribution size of particles freed from a catalytic converter, it was shown that large particles (>10 μm) dominate, reaching 62–67% of the total number of emitted particles, medium-size particles (3.1–10 μm) constitute 21%, and small ones (<3.1 μm) 13% (Artelt et al. 1999b; Ravindra et al. 2004). Particles with a diameter of >10 μm have the highest Pt content (Alt et al. 1997; Moldovan et al. 2002).

The distribution size of particles emitted from the converter (carrier material) is independent of the wear of the device.

The PGE emission rate and emission proportions among Pt/Pd/Rh depend on the following parameters (Artelt et al. 1999b; Whiteley and Murray 2003; Hoppstock and Sures 2004): speed of the vehicle, driving style (erratic stop–start flows), engine type, type and age of the catalytic converter, and nature of additives used in the fuel. Emissions may increase because of poor engine performance, for example, erratic fuel ignition or excessively hot exhaust fumes, which can damage the catalytic converter.

In practice, either direct or indirect approaches are used to determine the amount of PGEs emitted from a VEC (Palacios et al. 2000b; Moldovan et al. 1999).

The indirect technique consists of determining traffic-related PGE emissions in various environmental compartments, and comparing the obtained results with traffic volume data for vehicles equipped with catalysts (Helmers and Mergel 1998; Leœniewska et al. 2001). In the direct method, samples of exhaust fumes are collected that appropriately represent released particle levels. Sampling can be conducted under laboratory conditions by employing a computer-operated dynamometer. Dynamometric experiments, aimed at direct measurement of Pt from three-way catalysts, were performed on various engines and catalysts (Artelt et al. 1999b). The following parameters were considered while performing these experiments: manufacturer of the catalyst, catalyst wear, cubic engine capacity (displacement), and vehicle speed. The results (Artelt et al. 1999b) give rise to the following conclusions:

- Pt concentrations in exhaust fumes range from 7 to 123 ng m^{-3}, which is equivalent to the emission coefficient of 9–124 ng Pt km^{-1}.
- Similar emission levels were observed in catalytic converters of different designs and manufactured by various producers.
- The amount of released Pt increases with rising simulated speed, reaching values above 90 ng km^{-1} at 130 km hr^{-1} (for a new converter), and exhaust temperature.
- A three-way catalytic converter, in a car with a 1.4 L engine, shows a fourfold decrease in emitted Pt when compared to a car with a 1.8 L engine.
- No significant differences in the amount of released Pt were found when three-way catalytic converters (at various stages of wear) were compared at a speed of 80 km hr^{-1}.
- The amount of released Pt decreases with increasing catalytic converter wear; this phenomenon was observed at higher velocities.
- Pt emissions were higher in city driving, when compared to driving with a mean speed of ~80 km hr^{-1}, despite the fact that city driving is characterized as having lower velocity and lower exhaust fume temperature.

A similar experiment was performed by Moldovan et al. (1999). The results also demonstrated that PGEs are released at ng km^{-1} levels. Moreover, significant amounts of Pd and Rh were found in a diesel catalyst labeled as Pt only. Similarly, Pd and Rh were also found in exhaust fumes from these catalysts.

The first studies on Pt release from catalytic converters demonstrated that Pt^{4+}-containing oxide is produced from the reaction between metallic Pt and oxygen, or air, at 500 C. Until recently, it was concluded that the majority of Pt released from catalytic converters is in metallic form, and only a small percentage is oxidized, probably to the Pt^{4+} form (Artelt et al. 2000). The latest investigations have demonstrated that the soluble fraction of PGEs, emitted from automobile catalyst, may be higher than formerly thought (about 0–30% of the total PGEs released) (Moldovan et al. 2002). Palacios et al. (2000b) demonstrated that the soluble fraction of emitted PGEs constitutes 10% of the total, and exists mainly in the oxidized state and in chloride forms. It has also been proven that aluminum oxide and silica act as carriers for Pd released from catalytic converters, and that this metal probably occurs in halogenated form (Ravindra et al. 2004). Moreover, research on the water solubility of emitted Pt compounds, in samples collected from roads, undermines the hypothesis that PGEs are emitted from catalytic converters mainly in metallic form (Jarvis et al. 2001). Fig. 2 presents the content of different forms of Pt, Pd, and Rh that exist in exhaust fumes, and also presents the effect of engine age on emissions. Investigations concerning the determination of chemically active Pt in road tunnel dust have indicated that Pt is predominantly emitted from the converter in biologically available form (up to about 40%) (Fliegel et al. 2004).

4.2 *Emission of PGEs via Hospital Wastewater*

In recent yr, there has been a broad and increasing use of Pt complexes as anticancer drugs. Cisplatinum, oxaliplatinum, and carboplatinum are successfully used to treat selected malignant tumors. The above-mentioned anticancer drugs have been consecutively introduced since 1978. Research, including medical trials, is currently in progress on a new generation of Pt anticancer drugs (Esteban-Fernandez et al. 2007).

The biological half-life of the Pt anticancer drugs ranges from 160 to 720 d (Lenz et al. 2005). It has been estimated that Pt concentration in the urine of patients receiving chemotherapy is 40 times higher than normal physiological levels, even 8 yr after treatment ends (Schierl et al. 1995). The main accumulation sites for Pt in humans are kidneys, liver, spleen, and adrenal glands (Uozumi et al. 1993).

Approximately 70% of the total Pt dose received by a patient is excreted in urine. This amount includes 24–32% of cisplatin, 72–82% of carboplatinum, and 28–44% of oxaliplatin eliminated from the body, within 24 hr of treatment (Pyrzyñska 2000; Lenz et al. 2005). Pt excreted in urine reaches the sewerage system (Kümmerer and Helmers 1997), making hospitals an environmental emission source of this element. However, data from numerous studies indicate that hospitals play a secondary role in PGE emissions in comparison to other anthropogenic sources, particularly motor VECs (Kümmerer et al. 1999). The above data have been confirmed by investigations of Pt concentrations in municipal sewage;

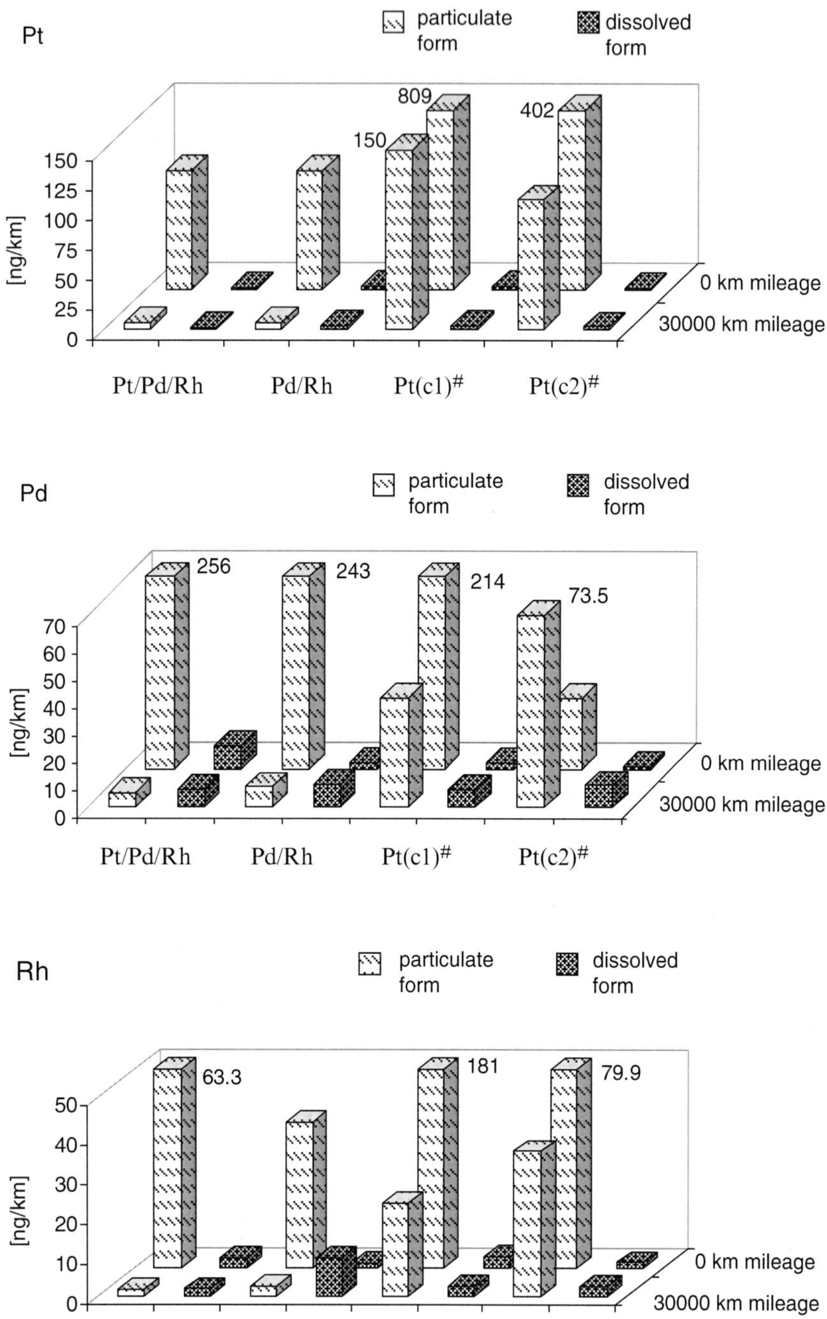

Fig. 2 Physical speciation of platinum group elements (PGEs) appearing in the exhaust gas from cars equipped with different types of catalytic converters. Data from new (0 km) and aged (30,000 km) engines are compared (Ravindra et al. 2004) #c1 and c2 are catalytic converters manufactured by the same producer (Palacios et al. 2000a, b)

the highest concentrations were measured at the beginning of rainy periods and at the end of winter (precisely when snow starts to melt). These findings suggest that the main emission source of Pt in municipal sewage is runoff from roads and urban areas (Kümmerer et al. 1999).

4.3 Other Environmental Emission Sources of PGEs

Analysis of PGEs in environmental samples do not support the thesis that emissions from VECs and hospitals constitute the total input of these metals into wastewater (Kümmerer et al. 1999). Therefore, it is necessary to consider other sources. The more important alternative sources are the electronics industry, glass production, jewelry production, dentistry use, the production of Pt-based pharmaceuticals, and industrial-scale catalytic processes.

The amount of emitted Pt found is influenced by sampling location, and residue levels largely depend on the nature of local emission sources (Lottermoser 1994). More information on PGEs emission sources and forms in which PGMs are emitted is needed to better understand the flow of PGEs in the environment (Moldovan 2007).

5 Bioavailability of and Occupational Exposure to PGEs

Only a limited number of studies have been published that deal with the bioavailability, mode of penetration into live organisms, and the environmental fate of PGEs. However, many scientists are increasingly concerned about the medium to high mobility of PGEs that are emitted to the environment from car catalytic converters. Moreover, Pt emitted in metallic or oxidized form (probably not strongly allergenic) may undergo transformation (when deposited as road dust) to a soluble form, and subsequently contaminate water, sediment, and soil, ultimately reaching the food chain and, thereby, animals and humans (Ducoulombier-Crepineau et al. 2007).

5.1 Effects on Plants

Certain PGE species can be remobilized as a consequence of being bound to soil particles that are taken up by plants, and thus may enter the food chain (Zereini et al. 1997; Hees et al. 1998; Lustig and Schramel 2000; Fritsche and Meisel 2004). Evidence indicates that PGEs, particularly Pd, are transported to biological materials by binding to sulfur-rich low molecular weight species in plant roots (Ek et al. 2004). Additionally, the solubility of PGEs may increase in the presence of natural complexing agents such as humic acids, as shown by Sures and Zimmermann. (2007). Schäfer et al. (1998) have performed an experiment in which the main result was a measurable transfer of PGEs from contaminated soil to plants. They concluded that Pt-, Rh-, and Pd-transfer coefficients are within the range of immobile to moderately mobile elements, such as Cu. The transfer coefficient decreases from Pd > Pt ≥ Rh, rendering Pd the most available element of this group. Data also show that there is uptake of the noble metals into different plant structures, in the following order: root > stem > leaf (Ballach and Wittig 1996; Schäfer et al. 1998; Messerschmidt et al. 1994).

5.2 *Effects on Animals*

Recent investigations with zebra mussels (*Dreissena polymorpha*) exposed to water, containing road dust or ground catalytic converter material, demonstrated that humic water from a bog lake clearly enhances the biological availability of particle-bound Pt; nonchlorinated tap water did not have this effect, and Pd showed the opposite result. No clear trend emerged for Rh. Differences in the effects of humic matter among the PGEs may be explained by the formation of metal complexes with different fractions of humic substances. The highest metal uptake rates and highest bioaccumulation plateaus were found for Pd, followed by Pt and Rh (Zimmermann et al. 2003, 2005; Sures et al. 2001). It was concluded that the uptake of Pt by the crustaceans *Asellus aquaticus* and *Gammarus pulex* is relatively high, when compared with other traffic-related heavy metals, or essential metals, such as zinc (Zn) (Haus et al. 2007). However, Microtox toxicity tests have shown that the EC_{50} (effective concentration of a substance that elicits 50% of a maximum response) of platinum chloride for the bacterium *Photobacterium phosphoreum* is 25 mg L^{-1}, which is much lower than for copper (200 mg L^{-1}) (Chen and Morrison 1994). A ranking of 80 metals for toxicity by Wolterbeek and Verburg (2001) revealed that Pt (II) and Pt (IV) ranked 10th and 11th, respectively [similar to Hg (mercury) (II) and Pb (lead) (IV)] (Wolterbeek and Verburg 2001; Sutherland 2003). PGEs, at high concentrations, cause water stress, chlorosis, and phytotoxicity in plants. Pt complex compounds are known to be mutagenic and carcinogenic, and inorganic Pt complex compounds are mutagenic in bacteria (Gebel et al. 1997; Sutherland 2003). The order in which PGEs are taken up is similar in exposed animals and plants: Pd > Pt > Rh. In animals, the liver and kidneys accumulate the highest levels of PGEs, especially Pd (Ek et al. 2004). Pt toxicity varies with oxidative state and electron configuration (Roshchin et al. 1984). The least harmful forms are metallic Pt and its nonsoluble salts (Table 2), although toxic effects are caused by a small group of Pt compounds that contain active substituents (chloride substituents are the most reactive). Studies carried out on rats showed that the amount of Pt retained in the body after 24 hr depended on the metal's chemical form; the following order was observed: $PtCl_4 > Pt(SO_4)_2 > PtO_2 > Pt$ (Moore et al. 1975a; Artelt et al. 1999a; Ravindra et al. 2004).

Table 2 Toxicity of specific platinum compounds in laboratory rats [after peroral (po) or intraperitoneal (ip) administration]

Platinum compound	Route of exposure	Toxicity LD_{50} (mg Pt kg^{-1})
Cis-platinum	Ip	7.4[a]
Sodium heksachloroplatinate (IV)	Po	15–50[b]
Hexachloroplatinic acid (IV)	Ip	15–19[b]
Platinum (IV) chloride	Ip	22[c]
Platinum (IV) chloride	Po	136[c]
Potassium tetrachloroplatinate (II)	Po	23–94[b]
Ammonium heksachloroplatinate (IV)	Po	88[b]
Platinum (II) chloride	Ip	490[c]
Platinum (II) chloride	po	>1400[c] >975[d]
Platinum (IV) oxide	po	>6900[c] >2926[d]

LD_{50} amount of a substance in mg kg^{-1} body weight, which causes the death of 50% (one half) of a group of test animals*LD* lethal dose
[a]WHO (1991)
[b]Lindell (1997)
[c]Holbrook (1976)
[d]Holbrook et al. (1975)

5.3 Evaluation of Human Health Risks

Health problems from occupational exposure to Pt were first reported in 1911 (Karasek and Karasek 1911; Farago et al. 1998). It is now well known that Pt compounds can cause a range of toxic effects in humans (Leœniewska et al. 2001). Miners and personnel employed in purifying and processing PGE-bearing ores are exposed to potentially high concentrations of these metals, which poses a serious health threat, including allergies to Pt salts. It is also known that certain Pt species have a skin-sensitizing potential, and that metallic Pd can cause contact dermatitis.

The first inhalation toxicity study from inhaling Pt-containing particles was published in the 1970s (Moore et al. 1975b; Ravindra et al. 2004). The level of soluble Pt compounds in air that can cause an allergic reaction is 0.1 μg m^{-3} (Rosner and Merget 2000). The following Pt compounds are considered to be the strongest allergens: hexachloro platinic acid (IV), ammonium and potassium hexachloroplatinate (IV), and sodium and potassium tetrachloroplatinate (II).

It has been established that Pt, and its derivatives, mainly accumulate at sites in the respiratory tract; however, significant Pt concentrations were also found in kidneys.

Breathing problems were reported for about one half of refinery employees exposed to the Pt salts used in industrial catalysts (Hughes 1980). The period between initial exposure to Pt salts and the first appearance of toxicity syndromes ranged from a couple of mon to 6 yr. In general, employees removed from the hazardous environment did not display the long-term effects of exposure (Ravindra et al. 2004).

Studies were also performed among employees of a catalyst manufacturing and recycling factory to evaluate exposure to metal compounds from the Pt group. The examination consisted of completing a work exposure and medical questionnaire, physical examination, skin prick test for the PGE compounds ($H_2[PtCl_6]$, $K_2[PtCl_4]$, $Na_2[PtCl_6]$, $IrCl_3$, $RhCl_3$, $PdCl_2$), and patch test for PGEs. Positive prick test reactions to PGE salts, at various concentrations, were found in 22 (14.4%) of 153 tested workers. Positive patch test reactions to Pt salts, at d 2, were seen in 2 of 153 subjects. The results of this study demonstrate that Pt salts are the most important allergens among PGEs compounds in the catalyst industry, and that clinical manifestations involve both the respiratory system and the skin. Hexachloroplatinic acid should be considered the most important salt to monitor, using skin prick tests. An allergic reaction was also observed to Ir and Rh salts in 2 of 22 workers (9%), who had positive patch test reactions (Cristaudo et al. 2005).

A 5 yr study, with annual monitoring, was also conducted in 159 catalyst production workers (94.6% of those recruited), 50 craftsmen (92.6% of those recruited), and 66 control subjects (76.7% of those recruited). Subjects were assigned to four exposure categories, and were segregated according to job title and location after the initial survey. The four categories were (1) high levels of Pt (n = 115), (2) persistently low levels of Pt (n = 51), (3) intermittently low levels of Pt (n = 61), or (4) no Pt (n = 48). Exposure assessment of airborne Pt and Pt in the serum of workers demonstrated clear differences between exposure categories. Smoking was a risk factor for Pt salt sensitization (in highly exposed subjects) (Merget et al. 2000). Calverley et al. (1995) also observed a definite association between smoking cigarettes and heightened Pt salt sensitivity. Risk of sensitization was about 8 times greater for smokers than for nonsmokers. The most typical allergy symptoms to Pt salts (Linnett 2005; Merget et al. 2000) were tearing, sneezing, rhinitis, shortness of breath, coughing, skin redness, itching or burning eyes, urticaria, and asthma.

Patients undergoing cancer treatment with Pt-containing medications, dental patients, as well as hospital personnel, dentists, dental technicians, and workers in drug manufacturing are exposed to PGEs (Sorsa and Anderson 1996; Nygren etal. 2002; Hann et al. 2005). The therapeutic use of cisplatin is often complicated by the occurrence of side effects (nephrotoxicity, severe nausea and vomiting, myelotoxicity, problems with hearing, and kidney diseases). Research has shown that some anticancer drugs may cause cancer (International Agency for Research on Cancer, IARC 1987, 1990). The cis-platin has been classified into the 2A group (probably carcinogenic to humans) by IARC.

There has also been an increased interest in the mutagenicity and potential carcinogenicity of Pt-containing drugs, particularly in relation to on-job exposure. Medical personnel are occupationally exposed as a result of their constant contact with and the increasing use of anticancer medications. Several reports have been published that describe the negative effects of such drugs on the health of personnel in oncological wards. The observed effects include hair loss, an increase in miscarriages, and abnormal fetal development (Sorsa and Anderson 1996). Some studies have shown elevated levels of Pt in blood or urine in graduate and staff nurses, compared with control subjects (Nygren and Lundgren 1997).

It has not yet been established whether inhalation of PGEs other than Pt causes more pronounced allergic reactions than are caused by Pt salts (Cristaudo et al. 2005). However, it was determined that direct contact of metallic Pd to skin may induce inflammation, which does not occur with metallic Pt. This effect has no consequence for the allergenic potential of Pd, or its compounds, which are absorbed via the respiratory tract. Notwithstanding, recent studies indicate that the allergenic potential of Pd compounds may have been underestimated (Schuster et al. 2000). A case of Ir allergies from occupational exposure was also reported (Merget and Rosner 2001; Ravindra et al. 2004; Cristaudo et al. 2005).

Pt was found not only in the blood and urine of occupationally exposed persons, but also in the body fluids of individuals who were not occupationally exposed. Average levels of 0.6 ng L^{-1} were found in blood of people living in Sydney, Australia (Vaughan and Florence 1992). Benemann et al. (2005) tried to identify and quantify the main exposure pathways of gold and Pt in the general adult population. Their studies were based on results obtained from the analysis of urine samples collected from 1,080 people aged 18–69. The results of these studies were that exposure correlated with the number of teeth that had noble metal dental alloy restorations.

A growing point of interest among scientists is the recent observation that small amounts of Pt have been found in silicone implants. Concerns have been raised that Pt may enter the body and cause adverse effects, by diffusing either through the intact implant shell or from an implant rupture. The recent data demonstrate that Pt leaks from intact implants and accumulates in lipid-rich fat or fibrous tissues (Maharaj 2004). Although the appearance of Pt residues in tissues is a good indicator for defective implants, there is no certainty that it causes health problems in women with implants (Flassbeck et al. 2003).

6 Concentration Levels of PGEs in the Environment

Scientists have warned that PGEs may negatively affect human health, not only from direct exposure via road dust and inhalation of airborne particles (about 30% of particles emitted from car catalytic converters are <10 μm in diameter, and are inhalable) (Gomez et al. 2001), but also through consumption of food and water (Rosner and Merget 2000; Artelt et al. 1999b). Therefore, it is important to define the human health risks associated with the presence of PGEs in the workplace; environmental monitoring of these elements at their emission sources is needed.

Bosch Ojeda and Sanchez Rojas (2007) have reported a trend toward PGEs increasingly being discovered in environmental samples. Rh, Pd, and Pt are known to accumulate in soils along roadsides. From the results of Morton et al. (2001), we clearly see that PGE concentrations in soils exposed to high traffic densities significantly exceed natural background values. The minor, but nonetheless significant, accumulation of Ir in soils is strong evidence of its increasing use in catalytic converters (Fritsche and Meisel 2004). Scientific results indicate that PGE concentrations in roadside soils are directly influenced by traffic conditions and are proportionate to distance from the road. The research of Morcelli et al. (2005) shows that the pattern of PGE residue distribution in roadside soil was similar to that of other traffic-related elements such as Zn and Cu; residues of Zn and Cu also decrease rapidly as distance from the traffic lanes increase. The work of Sutherland (2003) shows that roadside soils exhibited a significant decrease in Pt concentration with depth of soil sampling.

Very high concentration levels of PGMs reside in road and tunnel dusts. However, they are expected to rise further as the use of PGMs in catalytic converters grows. Leœniewska et al. (2004) conducted studies in polluted and nonpolluted areas, and produced results showing enhanced levels of Pt, Rh, and Pd in road and tunnel dust, or in grass growing near roads, compared to the geochemical background levels (samples collected from nonpolluted areas).

Schäfer and Puchelt (1998) demonstrated that PGMs emitted near roads accumulate in the vicinity, and are subject to transport by wind and washoff; subsequently PGM residues can be measured in airborne particulate matter and sediments.

Comparing the sampling yr 1993–1994 with 1995–1996, the latter showed a detectable increase in airborne Pt levels (Schierl and Fruhmann 1996). Schäfer and Puchelt (1998) revealed that people, residing particularly in urban areas, may be exposed to particles containing PGMs with enhanced reactivity because of cluster sizes in the nanometer range

The highest level of PGEs, in wetland sediment samples, was generally found at basin low points, and these concentrations may be related to the area of road surface drained and the traffic volume on proximate roads. Moreover, the results indicate that fractionation is occurring during transport through the drainage system, and suggest that a small portion of Pd in the road dust can be solubilized under natural conditions (Whiteley and Murray 2005).

Traffic-emitted PGEs can enter plants, under natural growing conditions, and show transfer coefficients similar to those of Cu, Ni, or Cd (cadmium), and much higher than those of Pb (Schäfer et al. 1998).

Pt was also detected in biota sampled from moderately polluted sites (Haus et al. 2007). More Pt was absorbed by *Asellus aquaticus* and *Gammarus pulex* than were other traffic-related heavy metals; uptake rates were more similar to those of essential metals such as Zn. By analyzing Sb (antimony) and Pt concentrations in tissues of those organisms, it could be shown that traffic-related heavy metals influenced most of the investigated aquatic. Pt was found in a majority of biota samples, and it emerged as being among the most widespread of the PGEs, even in moderately polluted water bodies.

Table 3 presents data on the PGE concentrations found in an array of environmental samples (Dubiella-Jackowska et al. 2007).

7 Summary

PGEs (Pt, Pd, Ru, Ir, and Os) are a relatively new group of anthropogenic pollutants. Specific useful properties of these metals (high resistance to chemical corrosion over a wide range of temperatures, high melting point, high mechanical resistance, and high plasticity) have fomented rapid growth of new and existing applications in various economic and industrial sectors. These metals are not only used in the chemical, petrochemical, electrical, and electronics industries but also

Table 3 Examples of PGE concentrations measured in environmental samples from various worldwide locations

Sample type	Sampling site	Analytes	Examples of concentration levels in environmental samples	Unit	Literature cited
1	2	3	4	5	6
Soil	Austria	Ru	0.01–5.77	ng g^{-1}	Fritsche and Meisel 2004
		Rh	0.03–13.2		
		Pd	0.29–24.5		
		Os	0.03–2.36		
		Ir	0.03–1.1		
		Pt	<LOD: 134		
	Brazil	Pt	0.31–17.4	ng g^{-1}	Morcelli et al. 2005
		Pd	1.1–58		
		Rh	0.07–8.2		
	Germany	Pt	<0.1–4	ng g^{-1}	Schäfer and Puchelt 1998
		Pd	<0.4–1.2		
		Rh	<0.1–0.6		
	Mexico	Pt	1.1–332.7	ng g^{-1}	Morton et al. 2001
		Pd	1.1–101		
		Rh	0.2–39.1		
	Australia	Pt	30.96–153	ng g^{-1}	Whiteley and Murray 2003
		Pd	13.79–108.45		
		Rh	3.47–26.55		
	Great Britain	Pt	<0.30–7.99	ng g^{-1}	Farago et al. 1996
Soil	Italy	Pt	1.8–4.7 (less polluted areas) 0.8–19.4 (urban areas)	ng g^{-1}	Cinti et al. 2002
	Italy	Pt	1.6–52 (urban areas)	ng g^{-1}	Cicchella et al. 2003
		Pd	8–110 (urban areas)		
Road tunnel dust	Poland	Pt	4.17–23.3 22.3 ± 4.4	ng g^{-1}	Leœniewska et al. 2004
		Pd	16.4 ± 1.9 3.1–23.9		
		Rh	6.76 ± 1.28		
Road dust	Poland	Pt	35.9–110.9 34.2–110.6	ng g^{-1}	Leœniewska et al. 2004
		Pd	42.2 ± 1.0 32.8 ± 3.8		
		Rh	10.8–19.7 6.0–19.5		
	Australia	Pt	53.84–419	ng g^{-1}	Whiteley and Murray 2003
		Pd	58.15–440.46		
		Rh	8.78–91.40		

(continued)

Table 3 (continued)

Sample type	Sampling site	Analytes	Examples of concentration levels in environmental samples	Unit	Literature cited
1	2	3	4	5	6
Road dust	Great Britain	Pt	0.35–32.7	ng g^{-1}	Farago et al. 1996
	Sweden	Pt	34.0–325.5	ng g^{-1}	Gomez et al. 2002
	Spain	Pd	70.8–202.7		
	Italy	Rh	4.8–101.8		
	Germany	Pt	135–303	ng g^{-1}	Djingova et al. 2003
		Pd	60–95		
		Rh	30–42		
		Ru	3.5–10.5		
		Ir	1.2–3.5		
	Scotland	Pt	1.8–335.1	ng g^{-1}	Higney et al. 2002
	Germany	Pt	101.3	ng g^{-1}	Sures et al. 2001
		Pd	21.3		
		Rh	18.7		
Airborne particulate matter	Czech Republic	Pt	9–62	pg m^{-3}	Vlasankova et al. 1999
			13–42		
			21–41		
		Pd	n.d.–280		
			n.d.–253		
			23–78		
Airborne particulate matter	Germany	Pt	<LOD: 43.1 samples collected 7.3 (mean) inside motor vehicles	pg m^{-3}	Schierl and Fruhmann 1996
	China	Pt	5.9–37.4	pg m^{-3}	Kan and Tanner 2005
			6.5–38.2		
	Germany	Pt	<7–62	pg m^{-3}	Schierl 2000
			21.5 (mean)		
Grass	Poland	Pt	8.98 ± 0.39	ng g^{-1}	Leœniewska et al. 2004
			8.27 ± 0.83		
		Pd	3.20 ± 0.23		
		Rh	0.68 ± 0.18		
			0.63 ± 0.07		
Dandelion	Germany	Pt	5.4–30	ng g^{-1}	Djingova et al. 2003
		Pd	0.83–1.5		
		Rh	2.0–7.0		
		Ru	0.4–0.9		
		Ir	<0.02–0.4		
Common plantain	Germany	Pt	3.6–10.1	ng g^{-1}	Djingova et al. 2003
		Pd	0.45–2.1		
		Rh	1.1–3.4		
		Ru	0.3–0.8		
		Ir	0.01–0.30		

(continued)

Table 3 (continued)

Sample type	Sampling site	Analytes	Examples of concentration levels in environmental samples	Unit	Literature cited
1	2	3	4	5	6
Fungi	Germany	Pt	5.9 ± 0.6	ng g^{-1}	Djingova et al. 2003
		Pd	0.2 ± 0.1		
		Rh	0.5 ± 0.1		
		Ru	0.25 ± 0.05		
		Ir	<0.004		
Darnel ryegrass (annual)			4.6–5.8		
			0.10 ± 03		
			2.1–2.2		
			0.4–0.6		
			0.1–0.2		
Moss			30 ± 2		
			2.4 ± 0.3		
			5.4 ± 0.5		
			0.9 ± 0.1		
			0.10 ± 0.03		
Tree bark	Great Britain	Pt	0.1–5.4 (mean = 1.5)	ng g^{-1}	Becker et al. 2000
		Pd	1.6–3.2 (mean = 2.1)		
		Rh	<0.05–1.8		
Urine	Germany	Pt	6.5 creatinine units (mean)	ng g^{-1}	Schierl 2000
Crustacean tissue (*Asellus aquaticus*)	Sweden	Pt	0.04–12.4 direct measurement	mg g^{-1}	Rauch and Morrison 1999
			0.16–4.5 measurement on sample washed for 24 hr		
Eel liver	Germany	Pd	0.18 ± 0.05 individual living in water polluted with street dust (10 kg dust/100 L water)	ng g^{-1} wtwt	Sures et al. 2001
Sediments from road runoff	Australia	Pt	9.03–103.80	ng g^{-1}	Whiteley and Murray 2005
		Pd	5.41–61.2		
		Rh	1.59–17.16		
Riverine sediments from road runoff	Great Britain	Pt	<0.29–4.42	ng g^{-1}	De Vos et al. 2002
		Pd	0.08–5.71		
		Rh	<0.11–0.26		
		Ru	<0.15–3.73		
		Ir	<0.03–2.69		

(continued)

Table 3 (continued)

Sample type	Sampling site	Analytes	Examples of concentration levels in environmental samples	Unit	Literature cited
1	2	3	4	5	6
Marine sediments	Pacific Ocean	Pt	2.7–22.8	ng g^{-1}	Terashima et al. 2002
		Pd	0.9–8.6		
	South Pacific	Pt	9.45–78.7	ng g^{-1}	Lee et al. 2003
		Pd	4.16–24.0		
		Ru	0.325–10.2		
		Os	0.033–0.809		
		Ir	0.269–13.8		
	North Pacific		117		
			90.1		
			27.2		
			1.08		
			12.7		
Primary effluent[a]	Germany	Pt		ng L^{-1}	Laschka and Nachtwey 1997
–October 1994			8.1–23.3 (dry season)		
			10.3–25.9 (rainy season)		
–June/July 1995			29.6–41.8 (dry season)		
			26.6–91.7 (rainy season)		
Secondary effluent					
–October 1994			4.4–7.6 (dry season)		
			8.2–10.7 (rainy season)		
–June/July 1995			6.4–9.2 (dry season)		
			11.7–12.4 (rainy season)		
Must	Germany	Pt	0.4	ng L^{-1}	Alt et al. 1997
Wine			0.5–2.4		
Hospital waste water	Germany	Pt	110–176 (d)	ng L^{-1}	Kümmerer and Helmers 1997
			38 (night)		
Snow	Russian Federation	Pt	<1–650	ng L^{-1}	Gregurek et al. 1999
		Pd	<1–2770		
		Rh	<0.5–19.0		

LOD limit of detection, *n.d.* not detected
[a]The cited work describes two sewage treatment stages, called primary and secondary treatments

have prominent uses in glass and jewelry production, and in the medical and dental sectors. Additionally, PGEs have exceptional catalytic properties, which render them particularly useful in vehicle and industrial catalytic devices. The growth of

PGE use, in various industries, has dramatically increased emissions of these metals to the environment; emissions from vehicle catalytic converters and hospital wastewater discharges are particularly significant.

The environmental benefits of using PGEs in vehicle catalytic converters are clear. These metals catalyze the conversion of toxic constituents of exhaust fumes (CO, HCs, NO_xs) to water, CO_2, and molecular nitrogen. As a result of adverse physico-chemical and mechanical influences on the catalyst surface, PGEs are released from this layer and are emitted into the environment in exhaust fumes. Research results indicate that the levels of such emissions are rather low (ng km^{-1}). However, recent data show that certain chemical forms of PGEs emitted from vehicles are, or may be, bioavailable. Hence, the potential for PGEs to bioaccumulate in different environmental compartments should be studied, and, if necessary, addressed.

The use of Pt in anticancerous drug preparations also contributes to environmental burdens. Pt, when administered as a drug, is excreted in a patient's urine and, as a consequence, has been observed in hospital and communal wastewater discharges.

Few studies have been published that address bioavailability, mode of penetration into live organisms, or environmental fate of PGEs. The toxic effect of these metals on living organisms, including humans, is still in dispute and incompletely elucidated. Contrary to some chlorine complexes of Pt, which most frequently cause allergic reactions, the metallic forms of PGEs are probably inert; however, they may undergo transformation to biologically available forms after release to the environment.

Because exposure to PGEs may result in health hazards, it is necessary to evaluate the risks of human exposure to these metals. Available data show that the highest exposed groups (Leœniewska et al. 2001) are individuals who work in refineries, chemical plants, electronics plants, jewelry production, oncological wards (medical personnel), and road maintenance; also highly exposed are women who have silicone breast implants. The effects of PGE exposure in live organisms include the following: asthma, miscarriage, nausea, hair loss, skin diseases, and, in humans, other serious health problems. As production and use of PGEs grow, there is a commensurate need to generate additional experimental and modeling data on them; such data would be designed to provide a better understanding of the environmental disposition and influence on human health of the PGEs.

Acknowledgment This work was supported by the Ministry of Science and Higher Education of Poland, Grant No. N N523 4410 33.

References

Alt F, Eschnauer HR, Mergler B, Messerschmidt J, Tölg G. (1997) A contribution to the ecology and enology of platinum. *Fresnius J Anal Chem* 357:1013–1019.

Artelt S, Creutzenberg O, Kock H, Levsen K, Nachtigall D, Heinrich U, Rühle T, Schlögl R. (1999a) Bioavailability of fine dispersed platinum as emitted from automotive catalytic converters: a model study. *Sci Total Environ* 228:219–242.

Artelt S, Kock H, König HP, Levsen K, Rosner G. (1999b) Engine dynamometer experiments: platinum emissions from differently aged three-way catalytic converters. *Atmos Environ* 33:3559–3567.

Artelt S, Levsen K, König HP, Rosner G. (2000) Engine test bench experiments to determine platinum emission from three-way catalytic converters. *In*: Zereni F, Alt F. (eds.) Anthropogenic platinum-group element emissions. Their impact on man and environment. Springer-Verlag, Berlin pp. 33–44.

Balcerzak M. (1997) Analytical methods for the determination of platinum in biological and environmental materials. *Analyst* 122:67R–74R.

Ballach HJ, Wittig GR. (1996) Reciprocal effects of platinum and Pb on the water household of poplar cuttings. *Environ Sci Pollut Res* 3:1–10.

Barefoot RR. (1999) Distribution and speciation of platinum group elements in environmental matrices. *Trends Anal Chem* 18(11):702–707.

Becker JS, Bellis D, Staton I, McLeod CW, Dombovari J, Becker JS. (2000) Determination of trace elements including platinum in tree bark by ICP mass spectrometry. *Fresenius J Anal Chem* 368:490–495.

Bencs L, Ravindra K, Van Grieken R. (2003) Methods for the determination of platinum group elements originating from the abrasion of automotive catalytic converters. *Spectrochim Acta B* 58:1723–1755.

Benemann J, Lehmann N, Bromen K, Marr A, Seiwert M, Schulz C, Jöckel K-H. (2005) Assessing contamination paths of the German adult population with gold and platinum. *The German Environmental Survey 2005. (GerES III). Int J Hyg Environ Heal* 208:499–508.

Bernardis FL, Grant RA, Sherrington DC. (2005) A review of methods of separation of the platinum-group metals through their chloro-complexes. *React Funct Polym* 65:205–217.

Bosch Ojeda C, Sanchez Rojas F. (2007) Determination of rhodium: since the origins until today ICP-OES and ICP-MS. *Talanta* 71:1–12.

Bradford CW. (1988) Platinum. *In* Seiler HG, Sigel H (eds.) Handbook of toxicity of inorganic compounds. Vol.10, Marcel Dekker, New York, Basel pp. 533–539.

Brook MA. (2006) Platinum in silicone breast implants. *Biomater* 27:3274–3286.

Calverley AE, Rees D, Dowdeswell RJ, Linnett PJ, Kielkowski D. (1995) Platinum salt sensitivity in refinery workers: incidence and effects of smoking and exposure. *Occup Environ Med* 52:661–666.

Chen W, Morrison GM. (1994) Platinum in road dusts and urban river sediments. *Sci Total Environ* 146–147:169–174.

Cicchella D, De Vivo B, Lima A. (2003) Palladium and platinum concentration in soils from the Napoli metropolitan area, Italy: possible effects of catalytic exhausts. *Sci Total Environ* 308:121–131.

Cinti D, Angelone M, Masi U, Cremisini C. (2002) Platinum levels in natural and urban soils from Rome and Latium. *(Italy): significance for pollution by automobile catalytic converter. Sci Total Environ* 293:47–57.

Cristaudo A, Sera F, Severino V, De Rocco M, Di Lella E, Picardo M. (2005) Occupational hypersensitivity to metal salts, including platinum, in the secondary industry. *Allergy* 60(2):159–164.

Cuif J-P, Blanchard G, Touret O, Marczi M, Quemere E. (1996) New generation of rare earth compounds for automotive catalysts. SAE Paper No. 961906.

Cuif J-P, Blanchard G, Touret O, Seigneuring A, Marczi M, Quemere E. (1997) (Ce, Zr)O_2 solid solutions for three-way catalysts. SAE Paper No. 970463.

De Vos E, Edwards SJ, McDonald I, Wray DS, Carey PJ. (2002) A baseline survey of the distribution and origin of platinum group elements in contemporary fluvial sediments of the Kentish Stour, England. *Appl Geochem* 17:1115–1121.

Djingova R, Kovacheva P, Wagner G, Markert B. (2003) Distribution of platinum group elements and other traffic related elements among different plants along some highways in Germany. *Sci Total Environ* 308:235–246.

Dubiella-Jackowska A, Polkowskaż ¯, Namiesnik J. (2007) Platinum group elements: a challenge for environmental analytics. *Pol J Environ Stud* 16:329–345.

Ducoulombier-Crepineau C, Feidt C, Rychen G. (2007) Platinum and palladium transfer to milk, organs and tissues after a single oral administration to lactating goats. *Chemosphere* 68:712–715.

Ek KH, Morrison GM, Rauch S. (2004) Environmental routes for platinum group elements to biological materials—a review. *Sci Total Environ* 334–335:21–38.

Energy Nexus Group. (2002) Technology characterization—reciprocating engines EPA, USA. http://wwwepagov/chp/documents/internal_combustionpdf.

Esteban-Fernandez D, Gomez-Gomez MM, Canas B, Verdaguer JM, Ramirez R, Palacios MA. (2007) Speciation analysis of platinum antitumoral drugs in impacted tissues. *Talanta* 72:768–773.

Farago ME, Kavanagh P, Blanks R, Kelly J, Kazantzis G, Thornton I, Simpson PR, Cook JM, Parry S, Hall GM. (1996) Platinum metal concentrations in urban road dust and soil in the United Kingdom. *Fresenius J Anal Chem* 354:660–663.

Farago ME, Kavanagh P, Blanks R, Kelly J, Kazantzis G, Thornton I, Simpson PR, Cook JM, Delves HT, Hall GEM. (1998) Platinum concentrations in urban road dust and soil, and in blood and urine in the United Kingdom. *Analyst* 123:451–454.

Flassbeck D, Pfleiderer B, Klemens P, Heumann KG, Eltze E, Hirner AV. (2003) Determination of siloxanes, silicon, and platinum in tissues of women with silicone gel-filled implants. *Anal Bioanal Chem* 375:356–362.

Fliegel D, Berner Z, Eckhardt D, Stüben D. (2004) New data on the mobility of Pt emitted from catalytic converters. *Anal Bioanal Chem* 379:131–136.

Fritsche J, Meisel T. (2004) Determination of anthropogenic input of Ru, Rh, Pd, Re, Os, Ir and Pt in soils along Austrian motorways by isotope dilution ICP-MS. *Sci Total Environ* 325:145–154.

Gebel T, Lantzsch H, Pleszow K, Dunkelberg H. (1997) Genotoxicity of platinum and palladium compounds in human and bacterial cells. *Mutation Res* 389:183–190.

Gomez B, Gomez M, Sanchez JL, Fernandez R, Palacios MA. (2001) Platinum and rhodium distribution in airborne particulate matter and road dust. *Sci Total Environ* 269:131–144.

Gomez B, Palacios MA, Gomez M, Sanchez JL, Morrison G, Rauch S, McLeod C, Ma R, Caroli S, Alimonti A, Petrucci F, Bocca B, Schramel P, Zischka M, Petterson C, Wass U. (2002) Levels and risk assessment for humans and ecosystems of platinum-group elements in the airborne particles and road dust of some European cities. *Sci Total Environ* 299:1–19.

Gregurek D, Melcher F, Niskavaara H, Pavlov VA, Reimann C, Stumpfl EF. (1999) Platinum-group elements. *(Rh, Pt, Pd) and Au distribution in snow samples from the Kola Peninsula, NW Russia. Atmos Environ* 33:3281–3290.

Hann S, Helmers E, Köllensperger G, Hoppstock K, Parry S, Rauch S, Rossbach M. (2005) Nuclear analytical methods for platinum group elements. ISBN 92-0-102405-3 International Atomic Energy Agency, Wien, http://www-pubiaeaorg/MTCD/publications/PDF/te_1443_webpdf.

Haus N, Zimmermann S, Wiegand J, Sures B. (2007) Occurrence of platinum and additional traffic related heavy metals in sediments and biota. *Chemosphere* 66:619–629.

Heck RM, Farrauto RJ. (2001) Automobile exhaust catalysts. *Appl Catal A* 221:443–457.

Hees T, Wenclawiak B, Lustig S, Schramel P, Schwarzer M, Schuster D, Verstraete D, Dams D, Helmers E. (1998) Distribution of platinum group elements (Pt, Pd, Rh) in environmental and clinical matrices: composition, analytical techniques and scientific outlook. *Environ Sci Pollut Res* 58:105–111.

Helmers E, Mergel N. (1998) Platinum and rhodium in a polluted environment: studying the emissions of automobile catalysts with emphasis on the application of CSV rhodium analysis. *Fresnius J Anal Chem* 362:522–528.

Higney E, Olive V, MacKenzie AB, Pulford ID. (2002) Isotope dilution ICP–MS analysis of platinum in road dusts from west central Scotland. *Appl Geochem* 17:1123–1129.

Holbrook DJ. (1976) Assessment of toxicity of automotive metallic emissions. Assessment of fuel additives emission toxicity via selected assays of nucleic and protein synthesis. Vol. 1, Research Triangle Park, N C, US EPA, Office of Research and Development, Health Effects Research Laboratories, (EPA/600/1-76/010a).

Holbrook DJ, Washington ME, Leake HB, Brubaker PE. (1975) Studies on the evaluation of the toxicity of various salts of lead, manganese, platinum, and palladium. *Environ Health Persp* 10:95–101.

Hoppstock K, Sures B. (2004) Platinum-group metals. In: Merian E, Anke M, Ihnat M, Stoeppler M. (eds.) Elements and their compounds in the environment. 2nd Ed., Wiley-VCH, Weinheim pp. 1047–1086.

Hughes EG. (1980) Medical surveillance of platinum refinery workers. *J Soc Occup Med* 30:27–30.

IARC. (1987) Monographs on the evaluation of carcinogenic risks to humans Suppl. 7, International Agency for Research on Cancer, Lyon p. 440.

IARC. (1990) Monographs on the evaluation of carcinogenic risks to humans. *Pharmaceutical drugs*. International Agency for Research on Cancer, Lyon, 50:47, 65, 123.

Jarvis KE, Parry SJ, Piper JM. (2001) Temporal and spatial studies of autocatalyst-derived platinum, rhodium and palladium and selected vehicle-derived trace elements in the environment. *Environ Sci Technol* 35:1031–1036.

Johnson Matthey. (2000) Platinum 2000. Interim review. Johnson Matthey Public Limited Company, London.

Johnson Matthey. (2007) Platinum 2007. Interim review. Johnson Matthey Public Limited Company, London.

Kan SF, Tanner PA. (2005) Platinum concentrations in ambient aerosol at a coastal site in South China. *Atmos Environ* 39:2625–2630.

Karasek SR, Karasek M. (1911) The use of platinum. Report of. (Illinois) Comission on Occupational Diseases to His Excellency Governor Charles S Deneen, Warner Printing Company, Chicago p. 97.

Kümmerer K, Helmers E. (1997) Hospital effluents as a source for platinum in the environment. *Sci Total Environ* 193:179–184.

Kümmerer K, Helmers E, Hubner P, Mascart G, Milandri M, Reinthaler F, Zwakenberg M. (1999) European hospitals as a source for platinum in the environment in comparison with other sources. *Sci Total Environ* 225:155–165.

Laschka D, Nachtwey M. (1997) Platinum in municipal sewage treatment plants. *Chemosphere* 34(8):1803–1812.

Lee C-TA, Wasserburg GJ, Kyte FT. (2003) Platinum-group elements (PGE) and rhenium in marine sediments across the Cretaceous–Tertiary boundary: constraints on Re-PGE transport in the marine environment. *Geochim Cosmochim Acta* 67(4):655–670.

Lenz K, Hann S, Koellensperger G, Stefanka Z, Stingeder G, Weissenbacher N, Mahnik SN, Fuerhacker M. (2005) Presence of cancerostatic platinum compounds in hospital wastewater and possible elimination by adsorption activated sludge. *Sci Total Environ* 345:141–152.

Leśniewska B, Pyrzyńska K, Godlewska-żyłkiewicz B. (2001) Platyna i jej związki w środowisku naturalnym człowieka- czy stanowią zagrożenie?. *(in Polish). Chemical News* 55(3–4):331–351.

Leśniewska B, Godlewska-żyłkiewicz B, Bocca B, Caimi S, Caroli S, Hulanicki A. (2004) Platinum, palladium and rhodium content in road dust, tunnel dust and common grass in Bialystok area. *(Poland): a pilot study. Sci Total Environ* 321:93–104.

Lindell B. (1997) Platinum. Vol. 14, DECOS and NEG Basis for an Occupational Standard ISBN 91–7045–420–5 ISSN 0346–7821, https://gupeaubguse/dspace/bitstream/2077/4161/1/ah(1997)_14pdf.

Linnett PJ. (2005) Concerns for asthma at pre-placement assessment and health surveillance in platinum refining–a personal approach. *Occup Med* 55(8):595–599.

Lottermoser BG. (1994) Gold and platinoids in sewage sludges. *Int J Environ Stud* 46:167–171.

Lustig S, Schramel P. (2000) Platinum bioaccumulation in plants overview of the situation for palladium rhodium. *In*: Zereni F, Alt F. (eds.) Anthropogenic platinum-group element emissions. Their impact on man environment. Springer-Verlag, Berlin pp. 95–104.

Maharaj SVM. (2004) Platinum concentration in silicone breast implant material and capsular tissue by ICP-MS. *Anal Bioanal Chem* 380:84–89.

Merget R, Rosner G. (2001) Evaluation of the health risk of platinum group metals emitted from automotive catalityc converters. *Sci Total Environ* 270:165–173.

Merget R, Kulzer R, Dierkes-Globisch A, Breitstadt R, Gebler A, Kniffka A, Artelt S, Koenig H-P, Alt F, Vormberg R, Baur X, Schultze-Werninghaus G. (2000) Exposure-effect relationship of platinum salt allergy in a catalyst production plant: conclusions from a 5-year prospective cohort study. *J Allergy Clin Immunol* 105(2):364–370.

Messerschmidt J, Alt F, Tolg G. (1994) Platinum species analysis in plant material by gel permeation chromatography. *Anal Chim Acta* 291:161–167.

Minakata K, Suzuki M, Nozawa H, Gonmori K, Watanabe K, Suzuki O. (2006) Platinum levels in various tissues of a patient who died 181 days after cisplatin overdosing determined by electrospray ionization mass spectrometry. *Forensic Toxicol* 24:83–87.

Mitzukami F, Malda K, Watanabe M, Masuda K, Sano T, Kuno K. (1991) Preparation of thermostable high-surface area aluminas and properties of the alumina supported Pt catalysts. Catalysis and Automotive Pollution Control II, Elsevier Science Publishers BV, New York pp. 557–568.

Moldovan M, Gomez MM, Palacios MA. (1999) Determination of platinum, rhodium and palladium in car exhaust fumes. *J Anal At Spectrom* 14:1163–1169.

Moldovan M, Palacios MA, Gomez MM, Morrison G, Rauch S, McLeod C, Ma R, Caroli S, Alimonti A, Petrucci F, Bocca B, Schramel P, Zischka M, Pettersson C, Wass U, Luna M, Saenz JC, Santamaria J. (2002) Environmental risk of particulate and soluble platinum group elements released from gasoline and diesel engine catalytic converters. *Sci Total Environ* 296:199–208.

Moldovan M. (2007) Origin and fate of platinum group elements in the environment. *Anal Bioanal Chem* 388:537–540.

Moore W, Malanchuk M, Crocker W, Stara JF. (1975a) Biological fate of a single administration of Pt-191 in rats following different routes of exposure. *Environ Res* 9:152–158.

Moore W, Malanchuk M, Crocker W, Hysell D, Cohen A, Stara JF. (1975b) Whole body retention in rats of different ^{191}Pt compounds following inhalation exposure. *Environ Health Persp* 12:35–39.

Morcelli CPR, Figueiredo AMG, Sarkis JES, Enzweiler J, Kakazu M, Sigolo JB. (2005) PGEs and other traffic-related elements in roadside soils from Sao Paulo, Brazil. *Sci Total Environ* 345:81–91.

Morton O, Puchelt H, Hernandez E, Lounejeva E. (2001) Traffic-related platinum group elements (PGE) in soils from Mexico City. *J Geochem Explor* 72:223–227.

Nygren O, Lundgren C (1997) Determination of platinum in workroom air and in blood and urine from nursing staff attending patients receiving cisplatin chemotherapy. *Int Arch Occup Environ Health* 70:209–214.

Nygren O, Gustavsson B, Strom L, Eriksson R, Jarneborn L, Friberg A (2002) Exposure to anticancer drugs during preparation and administration. *Investigations of an open and a closed system. J Environ Monit* 4:739–742.

Onovwiona HI, Ugursal VI (2006) Residential cogeneration systems: review of the current technology. *Renew Sustain Energy Rev* 10:389–431.

Palacios MA, Gomez M, Moldovan M, Morrison G, Rauch S, McLeod C, Ma R, Laserna J, Lucena P, Caroli S, Alimonti A, Petrucci F, Bocca B, Schramel P, Lustig S, Zischka M, Wass U, Stenbom B, Luna M, Saenz JC, Santamaria J, Torrens JM. (2000a) Platinum-group elements: quantification in collected exhaust fumes and studies of catalyst surfaces. *Sci Total Environ* 257:1–15.

Palacios MA, Gomez M, Moldovan M, Gomez B. (2000b) Assessment of environmental contamination risk by Pt, Rh and Pd from automobile catalyst. *Microchem J* 67:105–113.

Petrucci F, Violante N, Senofonte O, Cristaudo A, Di Gregorio M, Forte G, Alimonti A. (2005) Biomonitoring of a worker population exposed to platinum dust in a catalyst production plant. *Occup Environ Med* 62:27–33.

Peucker-Ehrenbrink B, Jahn B. (2001) Rhenium-osmium isotope systematics and platinum group element concentrations: loess and the upper continental crust. Vol. 2(10), Geochemistry, Geophysics, Geosystems, Paper No. GC000172.

Pyrzyñska K. (1998) Recent advances in solid-phase extraction of platinum and palladium. *Talanta* 47:841–848.

Pyrzyñska K. (2000) Monitoring of platinum in the environment. *J Environ Monit* 2(6):99N–103N.

Rao CRM, Reddi GS. (2000) Platinum group metals. *(PGM) occurrence, use and recent trends in their determination. Trends Anal Chem* 19:565–586.

Rauch S, Morrison GM. (1999) Platinum uptake by the freshwater isopod Asellus Aquaticus in urban rivers. *Sci Total Environ* 235:261–268.

Ravindra K, Bencs L, Van Grieken R. (2004) Platinum group elements in the environment and their health risk. *Sci Total Environ* 318:1–43.

Resano M, Garcia-Ruiz E, Belarra MA, Vanhaecke F, McIntosh KS. (2007) Solid sampling in the determination of precious metals at ultratrace levels. *Trends Anal Chem* 26:385–395.

Roshchin AV, Veselov VG, Panova AI. (1984) Industrial toxicology of metals of the platinum group. *J Hyg Epidemiol Microbiol Immunol* 28:17–24.

Rosner G, Merget R. (2000) Evaluation of the health risk of platinum emissions from automotive emission control catalysts. In F Zereniet al. (ed.) Anthropogenic Platinum Group Element Emissions. Springer Verlag, Berlin pp. 267–281.

Rudolph E, Hann S, Stingeder G, Reiter C. (2005) Ultra-trace analysis of platinum in human tissue samples. *Anal Bioanal Chem* 382:1500–1506.

Rühle T, Schneider H, Find J, Herein D, Pfänder N, Wild U, Schlögl R, Nachtigall D, Artelt S, Heinrich U. (1997) Preparation and characterization of Pt/Al2O3 aerosol precursors as model Pt-emissions from catalytic converters. *Appl Catal B* 14:69–84.

Schäfer J, Puchelt H. (1998) Platinum group metals (PGM) emitted from automobile catalytic converters and their distribution in roadside soils. *J Geochem Explor* 64:307–314.

Schäfer J, Hannker D, Eckhardt J-D, Stüben D. (1998) Uptake of traffic-related heavy metals and platinum group elements PGE by plants. *Sci Total Environ* 215:59–67.

Schierl R. (2000) Environmental monitoring of platinum in air and urine. *Microchem J* 67:245–248.

Schierl R, Fruhmann G. (1996) Airborne platinum concentrations in Munich city buses. *Sci Total Environ* 182:21–23.

Schierl R, Rohrer B, Hohnloser J. (1995) Long-term platinum excretion in patients treated with Cisplatin. *Cancer Chemother Pharmacol* 36:75–79.

Schuster M, Schwarzer M, Risse G. (2000) Determination of palladium in environmental samples. *In:*Zereini F, Alt F. (eds.) Anthropogenic platinum group element emissions their impact on man environment, Springer, Berlin pp. 173–182.

Shams K, Goodarzi F. (2006) Improved and selective platinum recovery from spent -alumina supported catalysts using pretreated anionic ion exchange resin. *J Haz Mat B* 131:229–237.

Sorsa M, Anderson D. (1996) Monitoring of occupational exposure to cytostatic anticancer agents. *Mut Res* 355:253–261.

Sures B, Zimmermann S, Messerschmidt J, von Bohlen A, Alt F. (2001) First report on the uptake of automobile catalyst emitted palladium by European eels. *(Anguilla anguilla) following experimental exposure to road dust. Environ Pollut* 113:341–345.

Sutherland RA. (2003) A first look at platinum in road-deposited sediments and roadside soils, Honolulu, Oahu, Hawaii. *Arch Environ Contam Toxicol* 44:430–436.

Terashima S, Mita N, Nakao S, Ishihara S. (2002) Platinum and palladium abudances in marine sediments and their geochemical behavior in marine environments. *Bull Geol Surv Japan* 53(11–12):725–747.

Twigg MV. (2007) Progress and future challenges in controlling automotive exhaust gas emissions. *Appl Catal B: Environ* 70:2–15.

Uozumi J, Ueda T, Yasumasu T, Koikawa Y, Naito S, Kumazawa J, Sueishi K. (1993) Platinum accumulation in the kidney and liver following chemotherapy with cisplatin in humans. *Int Urol Nephrol* 25(3):215–220.

Van den Tillaard JAA, Leyrer J, Eckhoff S, Lox ES. (1996) Effect of support oxide and noble metal precursor on the activity of automotive diesel catalysts. *Appl Catal B: Environ* 10:53–68.

Vaughan GT, Florence TM. (1992) Platinum in the human diet, blood, hair and excreta. *Sci Total Environ* 111:47–58.

Vlasankova R, Otruba V, Bendl J, Fisera M, Kanicky V. (1999) Preconcentration of platinum group metals on modified silicagel and their determination by inductively coupled plasma atomic emission spectrometry and inductively coupled plasma mass spectrometry in airborne particulates. *Talanta* 48:839–846.

Wedepohl KH. (1995) The composition of the continental crust. *Geochim Cosmochim Acta* 59:1217–1232.

Whiteley JD, Murray F. (2003) Anthropogenic platinum group element (Pt, Pd and Rh) concentrations in road dusts and roadside soils from Perth, Western Australia. *Sci Total Environ* 317:121–135.

Whiteley JD, Murray F. (2005) Autocatalyst-derived platinum, palladium and rhodium (PGE) in infiltration basin and wetland sediments receiving urban runoff. *Sci Total Environ* 341:199–209.

WHO. (World Health Organization) (1991) Environmental Health Criteria 125, Platinum. International Programme on Chemical Safety, Geneva, http://www.inchemorg/documents/ehc/ehc/ehc125htm.

Wolterbeek HT, Verburg TG. (2001) Predicting metal toxicity revisited: general properties vs. *specific effects. Sci Total Environ* 279:87–115.

Zereini F, Skerstupp B, Alt F, Helmers E, Urban H. (1997) Geochemical behaviour of platinum-group elements (PGE) in particulate emissions by automobile exhaust catalyst: experimental results and environmental investigations. *Sci Total Environ* 206:137–146.

Zimmermann S, Menzel CM, Stüben D, Taraschewski H, Sures B. (2003) Lipid solubility of the platinum group metals Pt, Pd and Rh in dependence on the presence of complexing agents. *Environ Pollut* 124:1–5.

Zimmermann S, Messerschmidt J, von Bohlen A, Sures B. (2005) Uptake and bioaccumulation of platinum group metals. *(Pd, Pt, Rh) from automobile catalytic converter materials by the zebra mussel (Dreissena polymorpha). Environ Res* 98:203–209.

Index

Q

R

S

Kolibri 360 GmbH
Ferdinand-Jühlke-Straße 7
99095 Erfurt, Deutschland
produktsicherheit@kolibri360.de